The Engineer's Contribution Contemporary Architecture

OWEN WILLIAMS

Series editors

Angus Macdonald
Remo Pedreschi

Department of Architecture
University of Edinburgh

To the memory of Philippa Cottam

The Engineer's Contribution to Contemporary Architecture

David Yeomans
David Cottam

Thomas Telford

Endorsed by

RIBA Publications

Published by Thomas Telford Publishing, Thomas Telford Ltd, 1 Heron Quay, London E14 4JD.
URL: http://www.thomastelford.com

Distributors for Thomas Telford books are
USA: ASCE Press, 1801 Alexander Bell Drive, Reston, VA 20191-4400, USA
Japan: Maruzen Co. Ltd, Book Department, 3–10 Nihonbashi 2-chome, Chuo-ku, Tokyo 103
Australia: DA Books and Journals, 648 Whitehorse Road, Mitcham 3132, Victoria

First published 2001

Also available from Thomas Telford Books

The Engineer's Contribution to Contemporary Architecture – Eladio Dieste. R. Pedreschi ISBN 0 7277 2772 9
The Engineer's Contribution to Contemporary Architecture – Anthony Hunt. A. Macdonald ISBN 0 7277 2769 9
The Engineer's Contribution to Contemporary Architecture – Heinz Isler. J. Chilton ISBN 0 7277 2878 4
The Engineer's Contribution to Contemporary Architecture – Peter Rice. A. Brown ISBN 0 7277 2770 2

The Architecture of Bridge Design. D. Bennett ISBN 0 7277 2529 7
An Introduction to Cable Roof Structures. Second edition. H. A. Buchholdt ISBN 0 7277 2624 2

A catalogue record for this book is available from the British Library

ISBN: 0 7277 3018 5

© David Yeomans, David Cottam and Thomas Telford Limited 2001

All rights, including translation, reserved. Except as permitted by the Copyright, Designs and Patents Act 1988, no part of this publication may be reproduced, stored in a retrieval system or transmitted in any form or by any means, electronic, mechanical, photocopying or otherwise, without the prior written permission of the Publishing Director, Thomas Telford Publishing, Thomas Telford Ltd, 1 Heron Quay, London E14 4JD.

This book is published on the understanding that the author is solely responsible for the statements made and opinions expressed in it and that its publication does not necessarily imply that such statements and/or opinions are or reflect the views or opinions of the publishers. While every effort has been made to ensure that the statements made and the opinions expressed in this publication provide a safe and accurate guide, no liability or responsibility can be accepted in this respect by the author or publishers.

Designed by Rob Norridge
Printed and bound in Great Britain by The Cromwell Press, Trowbridge, Wiltshire

Acknowledgements

A work that has had such a long development period as this one will have inevitably depended upon the generous help of a large number of people: including those who helped in the writing of the exhibition catalogue of 1986. Especial thanks are due to Owen Williams Consultants for access to and the use of their archive material, without which this book would have been impossible. If we single out a few more recent contributors here, others who may have contributed in the past but who do not find their name here will have to forgive us. We are grateful to Andrew Sayer for facilitating our recent use of the Owen Williams archive from which came the majority of illustrations in this book; to Peter Tuffrey of Ellis, Clarke and Gallennaugh for providing access to their Daily Express drawings and information on the development of this building; and to Phil Berridge of the University of Liverpool for his help with the preparation of illustrations. We would also like to thank Julia Elton and the late Frank Newby for making their collection of material available.

Preface

In 1986 the Architectural Association held an exhibition of the work of Owen Williams, which was accompanied by a catalogue of his work and a number of essays by one of the present authors and others[1]. The invitation to contribute to the present series, therefore, inevitably meant that the present work would draw heavily on the work already done. At the same time, it would be unnecessary to attempt either a complete biography of Owen Williams or a complete catalogue of his work as this already exists. The intention here is to assess his work within the general aims of the series, namely what did Owen Williams contribute to the architecture of his period? Restricting the book to this issue means only covering the broader aspects of Owen Williams' life and work in sufficient depth to place his architectural projects in context. However, it has been necessary to discuss some of the aspects of his bridge designs in detail because their development throws some light on the development of his thinking at the time. This was part of his preparation for his entry into architecture.

Owen Williams' work was covered extensively by the inter-war journals, especially in the pages of the *Architect and Building News*, but he also published himself and as a result his work became very well known in his day. Although his work was well known, he could not be described as part of any general architectural movement. Despite his work being admired by architects of the modern movement, he could hardly be considered a part of that circle. He refused to join the Modern Architectural Research (MARS) Group when invited to do so. His influence was to a large extent through a demonstration of what a functionalist approach to architecture could achieve. He also demonstrated the kind of contribution that an engineer could make to architecture, although in this he was not alone. Other engineers were to do much the same but not as architects in their own right. In this, Owen Williams is unique.

By tracing the development of his ideas, this work draws on a number of sources. Some of the early work was carried out as part of David Cottam's PhD thesis and for the exhibition catalogue already referred to. Some personal recollections were also used and some insight into Williams' own interests were provided by his own bound copies of *Engineering News Record,* which were acquired with the pieces of paper still in them marking the articles that he was interested in. But by far the most valuable source has been the collection of material held by the firm that he founded and which includes photographs of work in progress, working drawings, sketches and calculations.

As this book deals with buildings built entirely in the era of imperial units these have been used throughout.

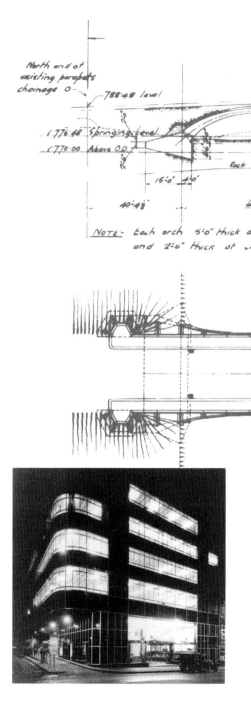

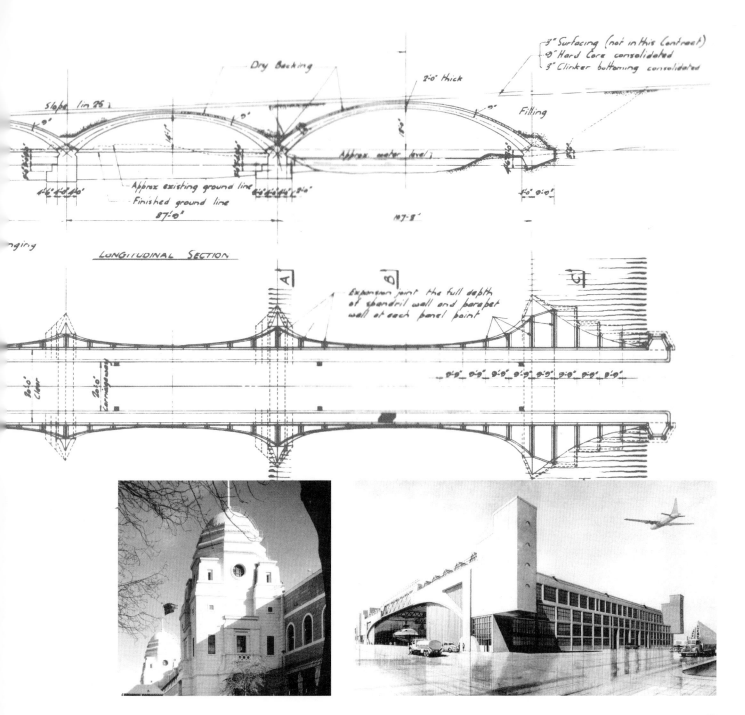

Contents

One	Owen Williams – The early years	11
Two	The British Empire Exhibition	19
	Critical comment	26
	Williams' and Ayrton's views	28
Three	Bridges and aesthetic theory	33
	Lea Valley viaduct	34
	Early A9 bridges	36
	Arch bridges	38
	Structural efficiency versus visual effect	39
	Wansford bridge	40
	Open spandrel arches	42
	Double cantilever bridges	44
	Bridge theory	46
	Ayrton's contribution	48
	Waterloo bridge	49
Four	The break	53
	The opportunities	57
Five	Dorchester Hotel and Daily Express	61
	Dorchester Hotel	62
	Daily Express	67
	Glazing	73
	Crediting the architect – structure and architecture	79
Six	Flat-slab buildings	81
	Factory planning	82
	Structure	87
	Glazing	90
	The result	92
	Sainsbury factory and warehouse	93
	The Pioneer Health Centre	95
	Postscript	105
Seven	Long-span buildings	107
	Empire Pool	108
	Odhams printing works (1935-38)	112
	Boots Drys	114
	BOAC hangar	117
	The plan of the main hangars	118
	The entrance door structures	121
	Wing hangars	122
Eight	Minor buildings	125
	Thurrock laboratories	126
	Stanmore flats	127
	Lilley and Skinner	128
	Dollis Hill synagogue	129
	Fire station	132
	Provincial newspaper	133
	Pistone laboratories	135
	Newspaper buildings	135
	Williams as an architect	137
Nine	Assessment	139
	Williams and concrete	140
	Williams as an architect	143
Endnotes		148
Selected List of Works		154
Bibliography		156
Index		158

Chapter One
Owen Williams – the early years

Chapter One
Owen Williams – the early years

The adoption of reinforced concrete as a commonly used architectural material at the beginning of the twentieth century naturally raised the issue in some minds of whether it would, or indeed should, lead to a new form of architecture. The question of an architecture of iron had already been explored as a theoretical issue in nineteenth-century France and, by the turn of the century, America was exploring the architectural possibilities of the skyscraper that had been facilitated by steel construction. It seems curious that Britain, which had embraced an arts and crafts architecture concerned with the expression of materials, does not seem to have taken much interest in these developments. Neither did the majority of the architectural profession in Britain show much interest in the possibilities of reinforced concrete as an architectural material. Its aficionados here during the inter-war years had to look largely to Continental Europe for examples of its expressive use and for the development of a new architecture based upon its unique properties. In Britain there were few attempts to explore this new material, so that by 1927 when Bennet and Yerbury published their book on concrete architecture they were able to include less than a handful of British examples and none that showed much grasp of the possibilities of the material.[1]

This is not the place to explore the reasons for this in any detail. Perhaps during the early 1920s there were few clients who would have been interested in this kind of architecture. Perhaps some of the reasons for this lay in the structure of the professions in Britain, for it seemed to require a rather closer collaboration between architects and engineers than was common here at the time.[2] Whatever the reason, some of the earliest mature essays in concrete architecture came from the work of Owen Williams who had trained as an engineer rather than as an architect. Moreover, he built these examples of concrete architecture while practising as an architect, rather than as an engineer in collaboration with an architect.

That Owen Williams was an enthusiast for concrete was hardly surprising. He had trained as an engineer at a time when reinforced concrete was beginning to make its mark as a structural material for building work and he spent his early years developing his skills in the design office of one of the leading reinforced concrete firms in the country. This was during the early years of the twentieth century, which saw major developments in reinforced concrete in Britain. When he left school in 1906 at the age of 16, Williams became an articled pupil to the Metropolitan Tramway Company, the engineering department of which was involved in the design of electric tramways and power stations. He was with them for five years and during that time he undertook part-time study in civil engineering and graduated in 1911 with a first-class honours degree. He then worked for two of the principal reinforced concrete companies in Britain, briefly for the Indented Bar and Concrete Engineering Company before going to one of their rivals, The Trussed Concrete Steel Company of England, where he remained for four years until 1916. This would have been his introduction to reinforced concrete design because the first course on the subject was not introduced until the year of his graduation.[3]

The market was then dominated by companies selling patent systems of reinforcement, mainly French or American-based because the early development of the material had been in these countries. Both the Indented Bar Company and the Trussed Concrete Steel Company were American in origin and both produced patented bar forms. With a wide variety of reinforcing types being produced there was no established simple method of reinforced concrete design, each type of bar required a different design method. Nor was there any standard specification for the concrete itself. Design for reinforced concrete structures was largely a service provided through the various contractors who were licensed to use the different products. Thus, construction in reinforced concrete was on a design and build basis. This was the commercial world that Owen Williams joined. In an attempt to free design from this commercial domination, the Royal Institute of British Architects (RIBA) set up a committee on reinforced concrete in 1907 and the Concrete Institute was formed in the following year.

The RIBA's committee had been established in 1906 partly at the suggestion of Dunn, co-author of the major British textbook on the subject, which had then just been published.[4] He pointed out the confusing state of affairs that existed with the various systems and methods of design, and he suggested that a more uniform approach was required. Initially, the committee considered that the arrangement by which design was normally entrusted to a specialist firm designing with one of the patent systems was probably desirable and so did not deal directly with design methods.[5] However, by 1911 the committee felt confident enough to set out rules for design in their second report, which eventually formed the basis of the London County Council (LCC) regulations for reinforced concrete published in 1913. This was, therefore, an interesting period in the development of reinforced concrete, coinciding with Williams' training in engineering and, as he was establishing himself in an engineering career with one of the reinforced concrete companies, these events could hardly have failed to influence someone with his lively mind.

Reinforced concrete was establishing itself during this pre-First World War period as an important material for commercial premises, warehouses and factories. Moreover, in the public sector it was being used for postal sorting offices under the direction of Sir William Tanner of the Ministry of Works who chaired the RIBA's committee. Nevertheless, it took a long time for consulting engineers to become established as a profession. Even allowing for the inevitable hiatus created by war-time conditions, when building work returned to normal in the inter-war period the practice of relying upon contractors to design the structure continued, in spite of concerns for the wastefulness of abortive design work to which this process inevitably led.[6] Nevertheless, there were a few important consulting engineers practising during the 1920s and events conspired to give Owen Williams his opportunity to enter this new world.

Williams' move to Trussed Concrete in 1912 had been an important one for him because this firm was important in introducing American techniques of reinforced concrete and industrial architecture into Britain. The firm marketed a patent type of reinforcing bar that combined tension reinforcement and shear reinforcement into a single bar. Also, the British director of the company who had come over from America promoted the use of reinforced concrete for factory construction, eventually publishing a book on this topic in 1917.[7] The firm was one of the more dynamic companies of the time, one of two that published a house journal to promote its products, and was to survive as a major contracting firm until after the Second World War. Unlike his contemporary, Oscar Faber, Williams did not have a university education that might have given him a broader perspective of engineering, so his training within the company would have been significant to the development of his ideas. Williams clearly played an important part in the company and its early development in Britain because within two years he became one of its chief designers and the influence of his period with the company can be seen in Williams' 1930s buildings.

Chapter One
Owen Williams – the early years

It is impossible to identify the projects with which Williams was associated during his twelve months with Indented Bar nor to say which structures he designed for Trussed Concrete. A review of contemporary journals suggest that one of the major projects then in Indented Bar's design office was a factory for Sainsbury's in Blackfriars, London,[8] a six-storey post and beam structure faced with conventionally designed brickwork that was typical of the time. Williams' autobiographical notes refer to many factory schemes at Trussed Concrete, particularly those erected for war purposes. His own calculations and photographs show his involvement with the Gramophone company building at Hayes, Middlesex (1913).[9] This factory, designed by Arthur Blomfield, was subsequently illustrated in Moritz Kahn's influential book on factory design.[10] The completed building was not untypical of other contemporary factory buildings although was perhaps higher than average. It was a six-storey post and beam structure with a water tower at one end that was some twenty feet above roof level. The concrete frame was exposed externally and the bays on the main façade were infilled with brickwork up to cill height with steel-framed glazing above. Monolithic concrete walls were used to fill the frames on the end walls, inset approximately six inches to maintain the expression of the frame.

As one of the chief designers at Trussed Concrete, Williams must have absorbed both their technology and their approach to design. As a commercial organisation producing buildings for industrial clients it was natural that efficiency in building design should be the hallmark of the Trussed Concrete Steel Company's image. During his time with this firm Williams must also have become acquainted with American developments. He would have seen the early issues of the company's quarterly journal *Kahncrete Engineering* (published from 1913) which reported exclusively on their work in Britain and America. Moreover, shortly after the war he began to subscribe to the American journal *Engineering News Record*, which regularly included articles on the development of concrete technology. By the time he left the Trussed Concrete Steel Company in 1917, he had worked on a number of war factories, including the development of a system of precast concrete.

He was then employed as an aircraft designer with the Wells Aviation Company for a little over twelve months, apparently designing a flying boat. Designing reinforced concrete frames for industrial buildings using an established method of design for a patent reinforcing bar system hardy allowed the exercise of much creative talent, so that frustration may explain what seems to be a rather curious move. However, his next appointment to a government-initiated research project into the development of concrete ships certainly did enable him to exercise his creativity in reinforced concrete design. For this project he led a syndicate in Poole, Dorset, which operated under the name General and Marine Concrete Construction (Williams System). This involved developing a novel form of construction and the means of analysing its behaviour, a problem that had to be approached from first principles and required both anlytical skills and his inventive turn of mind. Initially he seems to have been convinced of the usefulness of this project, filing a number of patents relating to concrete ship design. He also displayed his analytical approach to problems that he was later to display in relation to bridge design, writing a theoretical article on the economic size of concrete ships[11] and also publishing the results of experimental work that he had commissioned to examine the shear reinforcement of lightweight concrete that he was using for these hulls. His papers from this time include test reports from Kirkaldy's laboratories and photographs of ships under construction and at sea (*Fig. 1.1*). These were substantial craft but he eventually came to the conclusion that concrete ships were inferior to steel ones, mainly because of their poor resistance to impact and abrasion. Nevertheless, Williams was to return to this kind of work during the Second World War when he was involved with the design and construction of a number of Thames barges and two coasters, the latter of 2000 and 2500 tons.

The experience that he gained in this war-time work would have enabled him to develop his abilities in two areas that were to prove invaluable. The severe design constraints and the stringent requirements for efficiency in the use

Fig. 1.1. Williams' designed boat at sea.

15

Chapter One
Owen Williams – the early years

and economy of materials presented him with the opportunity to develop his design skills. The need to devise unique structural forms would have shown him the potential of reinforced concrete as a building material. Secondly, being leader of the project would have developed his organisational abilities and given him the experience and confidence to start his own firm once the war had ended.

After the war Williams returned to the commercial world of specialist contractors but this time as the proprietor of his own firm when, in 1919, he established Williams Concrete Structures Limited. This marketed his patent system of precast concrete construction known as 'Fabricrete',[12] which was used for industrial buildings. The design of this drew on his experience with the Trussed Concrete Steel Company because his system was very similar to that which he had helped to develop for them during the early part of the war and which was used for the Small Arms Factory in Birmingham.[13] Both were based on standardised precast concrete flared column heads and beams jointed together with in-situ concrete filling. Williams' new system differed only in the details of the jointing technique.

This was an appropriate method for the time. Precast concrete had been developed by a number of companies because a shortage of both timber and labour brought about by the war made it preferable to in-situ work. Following the war, the shortages may have eased but the capacity of the traditional building industry was still limited, thus encouraging the development of prefabricated building (especially for housing).[14] Therefore, precast concrete systems still provided a favourable solution for the industrialists who required new factory premises quickly. One of the factories built during the two-year life of the company was a tannery at Runcorn, the photographic records of which show a simple utilitarian structure devoid of any architectural pretensions. Illustrations of this building were included in many of the company's advertisements, so that it was presumably representative of the other buildings built using the same technique. Founding this company had been a natural step as the majority of successful concrete specialists in Britain were limited liability companies, rather than professional practices, promoting particular systems of concrete construction and reinforcement design. However, the direction of his career then changed in 1921 when he was appointed as consulting engineer for buildings of the British Empire Exhibition at Wembley.

In many ways this was a surprising appointment. There are no records to show how Williams acquired this job and there were other British engineers whom we might regard as equally well qualified. Among the alternatives that come to mind are, first, Oscar Faber, who had established himself as a specialist in reinforced concrete through both the research that he had carried out and his already published textbook on the subject.[15] However, Faber's early years in design were associated with the contractors Trollop and Colls and this might have disqualified him for a publicly funded project. Indeed, it even raises the possibility that seeing Williams get such a plumb job may have encouraged his own move into private practice in the same year. Another engineer who already had an established consulting engineering practice dealing with buildings was B. L. Hurst, although his building work up until then had largely been as consultant to contractors' design offices.[16]

Whatever the reason for the choice, this job provided Williams with a spectacular launch of his new career as a consulting engineer. Not only did his work for the British Empire Exhibition earn him a knighthood at the early age of 34 but, because of the extensive publicity which the buildings received, he acquired an unequalled reputation as one of Britain's leading reinforced concrete designers. Sir John Simpson and Partners were the architects for the Empire Exhibition but the job architect, with whom Williams worked, was Maxwell Ayrton. This was to be the beginning of a close working relationship between the two men that had the greatest influence on Williams' career. This began his interest in architectural design because, in their collaboration on this project Williams and Ayrton had the stated intention of developing reinforced concrete as a visually attractive material. They wished to develop the aesthetic possibilities of concrete and change its general association in people's minds with cheap, utility engineering and industrial structures. At the outset both believed that this could only be achieved through the effective collaboration of architect and engineer. Each brought to their joint work distinctive approaches to design. Williams the engineer, trained within the commercial pressures of concrete specialist firms, placed his highest priority on achieving economy of means while Ayrton, a traditionally trained architect, directed most of his attention to the architectural treatment of concrete. The way in which their collaboration developed can be seen first in the Empire Exhibition buildings and then in their bridge designs. Ultimately, as we shall see, the collaboration failed to develop the concrete architecture that Williams was looking for. The result was that Williams became disillusioned and eventually decided to work as an architect himself.

Chapter Two
The British Empire Exhibition

Chapter Two
The British Empire Exhibition

The British Empire Exhibition was first suggested by Lord Strathcona in 1913 but a decision was not taken to stage the event until after the First World War. Then it was seen as a reassuring public display of the strength of the British Empire in the aftermath of the war.[17] Each of the colonies was to be represented by separate buildings ranked in order of size according to their relative importance. Great Britain herself was to be represented by three major buildings, the palaces of Engineering, Industry and Arts, all to be temporary although the Palace of Industry and the façade of the Palace of Arts remain at the time of writing. The Palace of Engineering was only demolished in 1970. The only permanent structure was to be a new national sports stadium, designed to accommodate 125 000 people. There appear to have been four reasons for choosing concrete as the principal construction material:

- cost
- to ensure a short construction period of approximately twelve months
- because concrete was, rather curiously, considered as appropriate for temporary buildings, and
- to display the advanced state of British development of concrete technology.

For this last reason a British engineer would have to be chosen instead of one of the foreign-owned companies, such as Mouchel and Partners or the Trussed Concrete Steel Company.

The site chosen was a triangular, tree covered, 225 acres at Wembley and design work began in 1921. Ayrton's plan was much as one might expect of the time, a strong central axis leading up the hill to the stadium at the top that dominated the site. The north entrance was close to Wembley Park station (Bakerloo and Metropolitan lines) although a special station was also built (behind India's pavilion) served by the main line from Marylebone station. From the north entrance visitors walked up the broad central avenue flanked by the largest of the pavilions, the Palace of Industry and the Palace of Engineering. The Palace of Arts was much smaller and tucked round the back of the Palace of Industry. It was in fact on the cross axis that ran between the pavilions for New Zealand at the west end and India, appropriately, at the east end. Two major pavilions lay on this transverse axis, that for Australia (on Commonwealth Way) and that for Canada (on Dominion Way), while along it also lay the ornamental lake. As visitors walked up the main axis they would have had a view of the stadium that stood at the top and so dominated the site.

Williams does not seem to have been aware that the site planning might have been determined by such Beaux-Arts principles of symmetry, nor by any other aesthetic principles. In the lectures that he gave after the opening of the exhibition he emphasised the engineering considerations, claiming, for example, that the position of the stadium was chosen because its elevated position provided the best location for drainage and that the artificial lake at the foot of the hill had been created primarily as a balancing pool for rainwater run-off.[18] While these factors were clearly important aspects of the site planning and undoubtedly formed part of Williams' contribution to the design, he seems not to have paid much attention to the happy coincidence of these functional issues and the aesthetics of the site.

The Beaux-Arts strategy applied by Ayrton to the overall layout of the exhibition was also reflected in his treatment of the individual buildings. For each of these, with the exception of the smallest Oriental pavilions, in which neither he nor Williams had any involvement, his façades show a distinctly neo-classical idiom even though they are built of concrete. They were in fact merely screens round what were otherwise extensive pitched roofed sheds. The entrances to the two main pavilions were formed as simple recessed porches with their pediments supported on massive rectangular columns and flanked by large pylons (Fig. 2.1). The 'capitols' of these columns, no wider than their shafts, were simply indicated by recesses in the concrete. The Palace of Arts had an unashamedly neo-classical portico directly translated into concrete. Perhaps this was thought appropriate for its function.

Until that time there had been little concern for the aesthetic treatment of concrete. There had been some nineteenth-century essays on the use of the material, sufficient that in 1871 Arthur Blomfield published views on concrete as a material for architectural deign in the RIBA's sessional papers.[19] Many of the ideas which he expounded, for example, the use of shutter markings as part of the aesthetic treatment, and maximising the visual impact of daywork and expansion joints, seem, in retrospect, to be fairly self-evident opportunities presented by the material. Attitudes in Britain towards the use of concrete as an architectural material in the early years of the century have been discussed by Peter Collins, who records that there was hardly general enthusiasm for its use; pessimism is the term that he uses.[20] While there may have been considerable technical development by structural engineers in the early years of the century, there was little development by architects. The editor of *The Builder* appeared to be rather coolly disposed towards the material,[21] although there was one technical article on surface treatment of concrete that was based upon American experience which would have been useful.[22] Also, as one might expect, there was an article in the first

Fig. 2.1. Empire Exhibition — entrance to one of the British pavilions.

volume of *Kahncrete Engineering* on the same topic, although this would have had limited circulation.[23] In the same year as the article in *The Builder*, Professor Beresford Pite, at a meeting of the Concrete Institute, gave the first of two lectures that he was to give on the subject. This, by all accounts, seems to have been rather heavy going and according to one report seems to have had a limited audience. However, it could not have been of much practical use because, as Collins observes, '*Pite had a clearer idea of what to avoid rather*

Chapter Two
The British Empire Exhibition

than what to create.[24] Collins also notes that both Pite and Lethaby, who, lecturing two years later, was similarly unimpressed by concrete, were both speaking from a theoretical point of view, neither having any practical experience with concrete at that time. The most enthusiastic architect of this early period seems to have been C. F. A. Voysey who also gave a lecture for the Concrete Institute.[25] In this he raised the fundamental question about whether a new form of construction implied a new architecture and, if so, how were its rules to be determined?

Given the circumstances of the exhibition buildings, Ayrton was hardly likely to attempt to produce some radical new form of architecture with the new material: the kind of thing that had been attempted in Paris around the turn of the century. If there were any useful precedents for what he might do they were surely in the rather restrained approaches of Perret in his Théâtre des Champs-Elysées or in the postal sorting offices of Sir Henry Tanner, that had been rather curiously ignored by the architectural press. The later used simple pilasters and architraves to give some modelling to the surface. The aesthetics of concrete was here seen largely in terms of its surface treatment and this is clearly how Ayrton also saw it. He was against the idea of surface treatments that relied upon working

Fig. 2.2. Detail of the fluted rustication on the Wembley Stadium.

Fig. 2.3. Doorway in at the base of one of Wembley Stadium's towers.

the surface after striking the shutters, such as bush hammering, and what was developed at the Empire Exhibition was a rusticated appearance with vertical fluting and deep horizontal 'joints', the latter designed to conceal the daywork joints.

Of course we do not know what contribution the various parties may have made to the details of this design. As Ayrton had never used concrete before, at least not on this scale, he must surely have been obliged to collaborate with Williams and the contractor, McAlpine, in the design process. However, there can be no doubt that in the detailed handling of this rusticated treatment Ayrton showed considerable skill. A system of shuttering had to be produced that would give the effect of rustication; something of the jointing and the tooling effects of stone masonry in in-situ concrete. An appearance of rusticated masonry was used extensively in the base of the towers of the stadium by using fluting with recessed joints (*Fig. 2.2*), and this was carried through to fake vousoirs over the main doors traced out with this fluted appearance (*Fig. 2.3*). For the exhibition halls it was used much more discretely. Where there was to be the suggestion of columns at door and window openings the fluting was used to imply the presence of capitols and on long blank elevations, which were built of blockwork, a band of horizontal ribbing was formed into the edge of the precast blocks at the corners. Ayrton's approach was that concrete was simply a new form of masonry, translating this language of masonry into the surface treatment of the concrete.

The fact was that in their overall design Ayrton's buildings made no concessions to the use of concrete and this even extended to his architectural details. Indeed, in his traditional classical approach to the design of the façades he seems to have ignored both the practicalities of concrete and the real nature of the construction. There were deep window reveals looking just as if they were built into a thick masonry wall when in fact the whole was actually formed out of the thin concrete. The effect was that of a thick wall construction when in fact it was not, and this must have involved considerable complexity in the shuttering and would have presented problems in placing the concrete. In fact, examination of the concrete shows that these deep window reveals were cast separately onto the back of the wall once this had been cast. The recesses for the circular windows in the towers are actually thicker at the bottom than at the top suggesting that their weight was to be supported by this thickening. Ayrton was thinking simply in terms of masonry construction. Devising the means to achieve this kind of sham was presumably left to the contractor and the complexity of the shuttering may be imagined from the extent of the architectural details on the towers of the stadium (*Fig. 2.4*), details that are apparent in Ayrton's perspectives (*Fig. 2.5*). Of course, these details for the towers of the stadium are for the permanent building but he designed careful neo-classical details for the temporary buildings (*Fig. 2.6*).

In most of their buildings a clear distinction was made between the engineering and the architectural content, with Ayrton concentrating his attention on

Chapter Two
The British Empire Exhibition

the most visually important façades, leaving Williams to deal with the remainder. Williams would have had his hands full with the design of the large shed-like structures that formed the body of the exhibition halls, concealed behind these imposing façades and comprising either precast concrete frames or steel-trussed arches, and he used a variety of structures to roof the exhibition spaces. The Palace of Engineering resembled a giant factory with concrete columns and haunched beams carrying simple low pitched steel trusses, although the latter had slightly arched lower chords. The only relief to this rather dull structure was where the regular rhythm of the columns was broken by parabolic arches springing directly from the floor. The structure of the Palace of Industry was more interesting. The roof was of a steeper pitch and was carried by columns that branched into double cantilevers that carried principal rafters formed of open lattice trusses. Elsewhere this was all of precast concrete with the double cantilevers pierced for lightness, as were the precast arched principals that they carried.

While the treatment of the pavilions was rather like factory buildings with imposing fronts that screened the sheds behind, the stadium surely required a more holistic approach to its design. Precedents for this could be found in the Roman Coliseum and the more recent American stadia, a number of which were under construction at the time and were being reported in *Engineering News Record*.[26] However, it was treated in much the same way as the remainder. Its most important visual feature was the south façade that formed the termination of Wembley Way and Ayrton's first perspective drawing[27] of this (*Fig. 2.7*) shows his intention to produce a classical screen between the two domed towers.[28] Williams' role was to design the working details to produce this masonry effect. The sham that this led to is not only apparent in the surface treatment but also in the overall proportions and Williams' working drawings show the structural deceit he had to employ to produce the desired architectural effect. Ayrton's drawings of the wall panel between the two towers show excessively large columns, 6 ft x 2 ft and 4 ft x 2 ft in section at 13 ft and 17 ft centres, effectively the sizes that would be needed in a masonry structure. However, the working drawings show that these columns have little structural function. Each was hollow with an outside thickness of only 3–4 in. The main column supports were only 16 in. x 3 ft in section and were located in the centre of every third hollow column at more appropriate 40 ft centres.

This was typical of the approach used for the principal elevations of all the other buildings at Wembley. Nevertheless, in the remainder of the stadium Williams appears to have had a freer hand in adopting the most appropriate structural means for the functional requirements. The twin towers surmounting this main façade were a more sophisticated example of reinforced concrete design and Williams' drawings show how he borrowed from his experience of concrete ship design to produce a structure that accommodated Ayrton's simply drawn outline.[29] The domes themselves were designed like the hulls of Williams' concrete ships with a 3 in. thick shell reinforced by shallow curved ribs and tied by concrete cross members at their bases that also supported the foot of the flagpoles (*Fig. 2.8*). They were then supported on 4 in. walls stiffened by piers and buttressed by four concrete turrets.

For the remainder of the elevations, concrete walls were stiffened with solid 9 in. x 18 in. columns at 24 ft centres supporting parabolic arches that were simply repeated round the perimeter. These were occasionally interrupted by monolithic concrete stair towers providing access to the upper part of the open terrace. Unlike the principal elevations, these wall surfaces received little consideration and were left untreated from the plain shuttering. Internally Williams used a lattice steel framework to support the upper terraces, while the lower terraces were carried on earth excavated from the bowl. The composite nature of the stadium's

Fig. 2.4. Wembley Stadium towers.

Fig. 2.5. Ayrtons's perspective of a stadium tower.

Chapter Two
The British Empire Exhibition

Fig. 2.6. Neo-classical detail of the Palace of Industry.

structure supports Williams' insistence that, as engineer, he felt it necessary to use whichever structural materials and methods were best fitted to the particular problem. He explained this point in one of the lectures that he gave on the design of the exhibition:

'It has been my guiding principle to use no material where another material would better serve. To have used concrete where concrete should not be used would, apart from bad engineering, be of no service to the material itself.' [30]

What this does not say is that he was also borrowing ideas from other engineers of the time and some of the principal features of Wembley Stadium are similar to those of contemporary American stadia that were being described in *Engineering News Record* to which Williams was subscribing.[31] His decision to support the lower half of the sloped terraces on excavated earth received favourable comment at the time and, although a solution that he might well have arrived at by himself, it seems more likely that the idea came from its use in the Stamford Stadium at Washington University. This was one of those described in *Engineering News*.[32] Moreover, the same journal also described the use of a combination of lattice steel stanchions supporting upper concrete terraces at the Ohio Stadium, the type of composite construction that Williams used at Wembley.[33]

Critical comment

The treatment of these concrete buildings was no more than was expected at the time and although Williams' primary role was to translate essentially classically detailed masonry buildings into economical concrete forms, this collaboration of the disciplines received much favourable comment in the contemporary press. The buildings suggested that such a collaboration was something quite new. Reviewing the completed buildings in the *Architectural Review*, Harry Barnes wrote that:

'Theirs was the marriage of true minds to which there has been no impediment…They have affected something more lasting at Wembley — they have shown the possibility of collaboration and co-operation between architect and engineer, each enhancing the work of the other.' [34]

Other reviews made similar observations. For example, Lawrence Weaver in the same issue of the *Architectural Review* wrote:

'In the Wembley buildings…reinforced concrete has come into its own as a material used frankly and with vigorous invention for fine architecture.' [35]

This is hardly how we would view the buildings today and there was one dissenting voice at the time with which we might feel some sympathy. William Harvey, writing for the journal *Concrete* noted that:

'The classic tradition dies hard, and although one aspect of concrete work would class it with the continuous homogeneous walled and domed styles of the East, the facilities for shuttering straight solid masses have conspired with the recent revival of the Greek style to guide concrete design in the post and beam architecture of classic times.' [36]

His preference was for the stand of the Concrete Utilities Bureau designed by Clough Williams Ellis, which he thought showed more of the architectural possibilities of concrete.

Apart from William Harvey only one of the other critics may be regarded as well informed about reinforced concrete. This was Oscar Faber who reported on the construction of the exhibition for the *Architectural Review*[37] and who was less complementary than some other architectural critics. The surface treatment of the Palace of

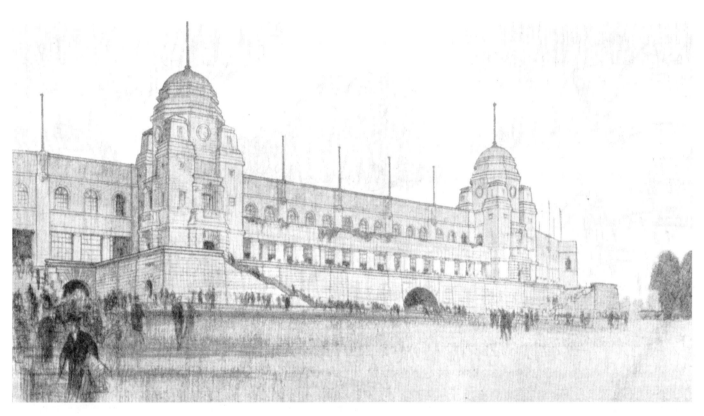

Fig. 2.7. Ayrton's perspective drawing of the stadium front.

Arts, whose development had been described by Ayrton in his RIBA lecture the year before, was much admired by Faber in his review but he was less happy about some other aspects of the concrete. He noted the poor workmanship with a good deal of honeycombing in the in situ. (His suggestion that this could have been avoided by the use of steel forms might not have been a practical proposition given the short construction period.) Of the stadium he said that: *'The inside was the roughest concrete job I had seen for some time'*. Preferring the steel trusses used on the Palace of Industry, he deprecated the use of concrete trusses on the Palace of Engineering. Here he noted all the common problems of this form of construction: large construction joints between the elements, their misalignment and damage in handling.

'The concrete roof trusses frequently have their

Chapter Two
The British Empire Exhibition

edges fractured off so as to show a line against the sky suggesting an attack by monster rats.'

Nevertheless, overall he considered it to be *'…a milestone on the road towards the proper treatment of concrete'*.

Whatever status reinforced concrete had enjoyed as an architectural material to date its extensive use in the Empire Exhibition meant that it had now arrived and, whatever faults there may have been in places, could hardly be ignored. Thus, when Beresford Pite spoke on its use at the RIBA in 1926 his views had rather changed from those he had expressed 15 years earlier.[38] Now, with the demonstration of its possibilities at Wembley, Pite seems to have been convinced of the future of the material and also considered the issue of a possible change in architecture that this would bring. In referring to the effect that its arrival would have on education he even went as far as saying *'what good will all the laborious study of the orders of stonework architecture be to the practitioner in ferro-concrete?'* We do not know what slides Pite may have used in his presentation but the record of the lecture was headed by a picture of Wembley Stadium. Thus, although nowhere referring directly to either that or the exhibition as a whole, there can be no doubt that his remarks were influenced by this building.

Pite certainly believed that concrete would produce a change in architecture. He noted the possibility of the use of cantilevers that concrete made possible. In this he was certainly prophetic because the idea was to be taken up in practice to such an extent that a few years later Goodhart-Rendel was to caricature its frequent and, to him, inappropriate use:

'…balconies thrust out as unconcernedly as though they were drawers pulled out of a piece of furniture. Everything seemed to stick by its edge to something else that was doubtfully secure in itself. Doubtfully secure, that is to say, to eyes accustomed to the forms that betoken security in masonry…' [39]

In his lecture, Pite opined that concrete also meant the death knell of the arch but here Owen Williams, who gave the vote of thanks, made a specific point of disagreeing with him (see chapter 3).

It must surely have been the building of the Empire Exhibition that led to the sudden interest in concrete among British architects that resulted in the *Architects Journal* having a special issue on concrete in 1926[40] and encouraged T. P. Bennet and F. R. Yerbury and their publisher to produce their book on concrete architecture in the following year.[41] This showed a wide range of material, from modern movement buildings in Europe to American buildings that varied between the neo-classical and the fanciful. Continental examples included Max Berg's Centenary Hall in Breslau (1912–13), which was particularly striking in having columns in the entrance porch that were far more slender than anything that could be achieved in masonry. It also had examples of the early work of Perret, Lucat and le Corbusier from France. Among the relatively few British examples were factory designs and Purley waterworks by Thomas Wallis, which was significant in view of later developments.

Williams' and Ayrton's views

Whatever effect the building of the Exhibition had on the general development of ideas about concrete architecture, of more importance are the views of Ayrton and Williams and the way that these affected their work together over the next few years. Unfortunately, there is no surviving correspondence between the two that might have shed some light on this and we have only their published articles that might explain the development of their ideas. Their public statements confirmed their joint commitment to developing concrete as a building material of quality but there were differences in their views that would eventually become significant. The first to appear after the Wembley Exhibition provide a clear statement of both Williams' and Ayrton's views about the need for close collaboration between architect and engineer if the concrete was to produce its own forms of architectural expression.

Both regarded the exhibition buildings as a landmark in the development of reinforced concrete. Ayrton did not see this as being achieved through the creation of new architectural forms (although he admitted that it had the potential to achieve this). In his lecture that he gave to the RIBA he declared that it was *'largely in the surface treatment of reinforced concrete that the architect will find his opportunity'*.[42] The exhibition buildings had proved that concrete could produce visually acceptable buildings. Previously, concrete had been regarded as a cheap material associated with the utilitarian designs of engineers. In justification of his own classical designs, Aryton maintained that architects were under an obligation not to produce revolutionary architectural forms that, by alienating public opinion, would be prejudicial to the material. He believed that concrete could best be developed as an architectural material through an evolutionary approach in which engineers and architects would have to collaborate. Engineers would provide the essential technological skills with architects producing acceptable surface finishes. Only in this way, he maintained, would concrete develop a distinctive style of its own.

What did Williams think of the structures that he had to use for Ayrton's classical façades? Despite Ayrton's architectural control over the design of the principal façades, during 1924 and 1925 Williams constantly referred in his lectures to 'economy of means' as the design principle that he had observed throughout.[43] This, he claimed, was the best tenet of all engineering design and any structure which failed in respect of this

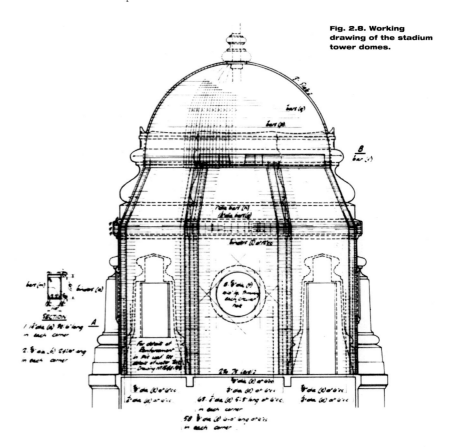

Fig. 2.8. Working drawing of the stadium tower domes.

Chapter Two
The British Empire Exhibition

principle was, in his view, bad engineering. Although he never directly criticised the 'misuse' of concrete in those parts of the exhibition buildings, many of his articles published between 1924 and 1925 contain veiled references to his disapproval. For example, in a 1924 lecture he justified the use of concrete in this way by claiming that architecture was some years behind engineering in its assimilation of the new structural techniques. He was hopeful, however, that through an evolutionary process architects and engineers working in close collaboration would be able to produce new forms of architecture that were more responsive to the unique qualities of the material:

'Architecturally [concrete] has largely been regarded as a somewhat mysterious alternative, only to be adopted for economy, or put in to support a façade of other materials in which it is hastily clothed to hide its nakedness. The buildings of the Exhibition witness a new architectural stage in the history of reinforced concrete. It is entering upon an era possibly of slow acceptance to architects, but ultimately must develop an architecture of its own, in the same way as it has developed a branch of engineering.' [44]

Under the heading 'Concrete as a Partnership of Engineering and Architecture', Williams saw a coming together of architecture and engineering. To him there was little difference between the disciplines of architecture and engineering, and he maintained that it was necessary for architects and engineers to recognise this and ultimately unite in a single profession if concrete was to achieve the high qualities associated with other materials. This sentiment is important in the light of Williams' future career so that it is worth quoting at some length.

'A considerable amount of time must elapse before the "concrete sense" can be acquired; that is to say, before any individual can achieve singly a complete and easy mastery of both the engineering and architectural technique. The engineer and the architect have a long road to travel before their separate roles can be played by one man. Till that end is achieved the fullest expression of concrete can not be attained. But the goal may be reached more quickly by sympathetic collaboration on both sides. The engineer must realize that sound architecture is only sound engineering and the architect must believe that sound engineering is the only sound architecture. Beauty of design must not be considered the sole property of the architect, nor must the engineer assume exclusive possession of the theories of stability. The eye of the architect may often be a more truthful guide than the slide rule of the engineer. On the other hand, the theories of the latter may achieve something more perfect than the architect can, because the engineer is in closer touch with the demands of the material.' [45]

It is clear from this that Williams regarded the unification of architectural and engineering abilities in one person as the ideal solution to developing a form of architecture that truthfully responded to the qualities of reinforced concrete. He had chosen to publish these ideas in a journal that could hardly have been read by many architects and, therefore, could have had little influence on the architectural thinking of the day. Nevertheless, he was not alone in this thinking because Goodhart Rendel was to write that the separation of architect and engineer:

'…is recent, regrettable and — it may be hoped — impermanent. The separation that is essential and irreconcilable is not between engineer and architect, but between engineering and architecture, not between men but between functions; functions wholly disparate, each of which, nevertheless, can often best be performed by the man who is also performing the other.' [46]

In contrast, Ayrton did not see any possibility of the eventual unification of the two disciplines, believing engineering to be:

'…a science which architects can never hope to take over entirely; it is far too complex and scientific a business to be thrown in as a side issue of architects' work. But its use should have this great benefit, that of bringing architects and engineers more closely together, a thing very urgently needed.' [47]

Instead he was rather defensive, seeing engineers as potential competitors. Thus, for him, collaboration was to some extent a defence against this possibility. In one discussion at the RIBA he warned his professional colleagues that if they did not begin to work more closely with engineers, developing architectural forms responsive to the new materials, they would be in danger of losing a large amount of their workload to the engineering profession. It was collaboration that he sought.

'The close union between these two great professions is one of the benefits which should transpire from the coming general use of reinforced concrete. The practice of working separately has unfortunately been too general.' [48]

He believed that architects could contribute most to this collaboration by concentrating on the aesthetic appeal of concrete, particularly in relation to its surface finishes:

'As a material used by engineers, the surface treatment is not one which they have to consider very seriously…it is largely in the surface treatment of reinforced concrete that the architect will find his opportunity.' [49]

Williams appeared to agree with Ayrton about the architect's role in developing the aesthetic qualities of concrete but the question raised is what he meant by this. In an unpublished and, unfortunately, undated manuscript in his private papers is the following passage:

'For this material architects are under an obligation to engineers and they must pay their debt by the study of its aesthetic qualities and do for it what their predecessors did in the past for the ancient materials of construction.' [50]

Perhaps it was tact that prevented Williams from voicing this view publicly because the implication is that architects had neglected the development of this material and now found themselves in a subservient role.

Both were interested in developing concrete as a building material of quality and both were convinced of the need for collaboration to effect this objective. But we should note here Ayrton's principal concern with surface finishes. Nevertheless the continuation of this relationship must surely have been Williams' initiative because it can have been no coincidence that they were both appointed to work on a series of Scottish bridges for the Ministry of Transport in the years following. We must assume that when Williams was engaged to work on these he would have suggested Ayrton as the consulting architect.

Chapter Three
Bridges and aesthetic theory

Chapter Three
Bridges and aesthetic theory

It may seem strange to discuss bridge design in a work that sets out to assess Owen Williams' contribution to architecture, especially so since his later bridges, those for the first phase of the M1 motorway, received a rather mixed reception when they were built. But his bridge designs should not be judged on these rather heavy looking structures nor is it simply for their designs alone that we need to consider his earlier bridges. The significance of his early bridge designs, produced in collaboration with Ayrton in the years immediately after the Empire Exhibition, is that they are an important phase in the development of Owen Williams' aesthetic ideas. They represent the major body of work that they carried out together and, therefore, must have been important in shaping his attitudes to architecture and architects in general.

Their statements about the Empire Exhibition imply that Williams and Ayrton shared a similar objective in their mutual desire to develop the visual treatment of concrete and this was to form the basis of their work together on the bridges that they then designed. Williams presumably found Ayrton a sympathetic colleague because it must surely have been he who suggested the latter as consulting architect for the bridge work. Williams would have been the lead consultant for this work taking responsibility for the major structural decisions, a position

Fig. 3.1. Drawing showing Ayrton's design for the Lea Valley Bridge.

in which he would normally be determining the overall form. Ayrton was appointed as consulting architect, thus, one might assume, concentrating simply on the visual aspects of the design and, in particular, on the details of surface finish. So, compared with the Empire Exhibition, their working relationship would now have been reversed. Nevertheless, while the structural issues of bridge design must have given Williams a strong position in this collaboration, it appears to have taken some time before he developed sufficient confidence to be able to express his engineering ideas in built form while Ayrton's visual ideas are clearly evident in their early work together.

Of course, as with all collaborations, it is impossible to tell, without being present, who contributed what to the overall design. Sometimes this may not matter but what is fascinating here is the striking differences between some of the designs, differences that suggest a possible shift in their relationship that occurred with the development of Williams' ideas.

Lea Valley Viaduct

Their first bridge project on which they collaborated was the Lea Valley Viaduct and Bridge (1924), which formed part of the new North Circular Road, London. Perhaps 'collaborated' is rather overstating the nature of their relationship in this case because, if anything, it was even less successful than the exhibition buildings in integrating the architectural and engineering contributions to the design. Indeed, the demarcation between Williams' and Ayrton's contributions was such that they could have been undertaken as separate contracts. The essential engineering problem was to carry the road over a large expanse of low-lying ground, for the most part a shallow depression of only 16 ft. Williams' first drawings show a flat-slab structure but, in the event, he designed a flat concrete road deck supported on a series of reinforced concrete frames, each of which comprised four large tapered columns with downstand beams. Williams' structural solution was devoid of any decorative treatment or stylistic modifications.

Our common experience of structures of this kind, where there is no dramatic fall in the landscape, is that one is hardly aware of the existence of a bridge and this presented a problem for Ayrton. The fact that the viaduct was barely noticeable in its surroundings and was completely hidden from the road users' point of view seemed to offend his architectural instincts. As he later said in a paper he delivered to the RIBA:

'In many instances there is nothing to call ones attention to the fact that one is approaching or even crossing a bridge. I feel that this definitely justifies, in certain cases, some form of superstructure, as in the case of the Lea Valley Viaduct and Bridge over the Lea Navigation River…I am well aware that this can be defended solely on aesthetic grounds.' [51]

To resolve this problem, as he saw it, he added a pair of large monumental features flanking each approach to the viaduct (Fig. 3.1). These wholly redundant, independent structures consisted of high, curved concrete walls articulated by pilasters. Williams could have contributed nothing to their design although it was a drawing of these that was reproduced in an article on their partnership in the *Architects Journal*.[52] What they provide here is an indication of the differences between the approaches of Williams and Ayrton.

Williams' attitude at the time to these pieces of sculpture, for they can hardly be called anything else, is not recorded.

Fig. 3.2. Findhorn Bridge as constructed.

Chapter Three
Bridges and aesthetic theory

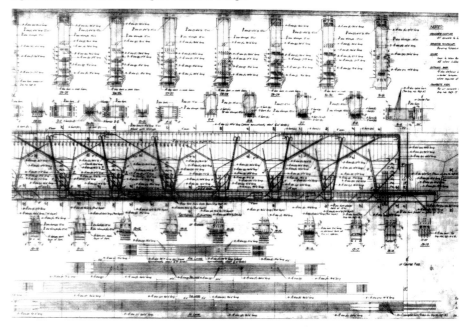

Fig. 3.3. Reinforcement drawing for Findhorn bridge.

However, it can be assumed from his later and more general pronouncements that he would not then have agreed to their inclusion. Indeed, he might well have used them to illustrate his argument that the objectives of engineer and architect were in direct conflict — the one committed to 'efficiency', the other 'effectiveness'.[53] 'Effectiveness' was the word that he used to refer to visual effect. If this was to be the nature of their collaboration and these pylons the only contribution that Ayrton made to the design their relationship might have been short-lived but he also helped to devise a simple modelled pattern in the concrete surfaces of the parapets. Ayrton advised that these receive a bush-hammered finish, an apparent change of mind from his earlier views about this kind of surface treatment, but it was a wise choice and has stood the test of time well. In fact, bush hammering was to be used on many of their subsequent bridges and was without doubt a positive contribution to their appearance.

Early A9 bridges
The pair then collaborated on a number of bridges in Scotland, a variety of types designed to cope with a wide range of different situations. Forty bridges were required to form part of the A9 road reconstruction between Perth and Inverness, the most important of which were assigned to Williams and Ayrton by the Ministry of Transport.[55] In these we see clear examples of how aesthetic ideas could shape not just the details of the structure but its overall form. In fact it was one of the first of these A9 bridges, the Findhorn Bridge at Tomatin (1925), that was perhaps their most successful scheme in terms of an integration of engineering and architectural solutions. Here, instead of Ayrton devising some unnecessary additions to draw attention to the bridge, a monumental effect was achieved within the overall engineering scheme and this was cleverly expressed in what we must assume was Ayrton's concrete detailing.

The bridge comprised two long spans, each of 98 ft, with a central support. One span was over the river itself while the other was over a wide flood area to one side (Fig. 3.2). In addition to this there was a small three-arch stone bridge on the approach that had to be widened. The road deck of the main bridge is suspended from deep vierendeel girders, 38 ft apart, which provide the road users with an impressive arcade through which they drive. This would clearly have satisfied Ayrton's requirements

Fig. 3.4. Design perspective of the Findhorn Bridge.

for a clear expression of the crossing but there were also sound engineering reasons for this form of structure. The ground conditions were poor, which would have made it difficult to contain the thrusts of an arch bridge. The structure they used avoided this, as well as providing a visual feature, although more conventional tied arches would have had the same effect. Therefore, the choice of vierendeel suggests a genuine collaboration over choice of form and, although we have no direct evidence for it, we can imagine Williams putting alternatives to Ayrton. The arrangement of reinforcement within the vierendeel (*Fig. 3.3*) then suggested the size and shape of the polygonal openings within them and Ayrton used the shape of these to great advantage in the modelling of the concrete surfaces.

While this was the first convincing collaboration between the two, with the result so clearly achieving what was intended by the design drawings (*Fig. 3.4*), it was achieved at considerable cost. Although comparable with many equally successful structures in span and site conditions, it was by far the most expensive of the bridges on the A9 with a contract price of £33 146. A large part of this must surely be attributed to shuttering costs because the form of the vierendeel was far from a simple deep pierced beam. Instead, the general impression given is of a series of faceted arches rising from a level parapet. On the inside the parapet wall is vertical, and inset between the arch forms, while from outside these arch forms spring from a sloping surface. Moreover, below each is a bracket that appears to serve no function (unless they were supports for the shuttering). This theme was then carried through into the walls of the turrets at either end that marked the entrances to the bridge.

But it was not just the surfaces of the arcade that were carefully modelled. Openings in the supports were shaped in a manner whose subtlety cannot be

Chapter Three
Bridges and aesthetic theory

appreciated fully without standing under the bridge. And even the supports for the widening of the approach span were shaped with a care that can only be seen from close to (*Fig 3.5*). Thus, while Ayrton and Williams were beginning to integrate their design skills to produce a satisfactory combination of structural form and visual effect, they had not managed to combine this with an economical design.

Arch bridges

This rather extravagant use of concrete was repeated in three other bridge designs for the A9 before 1926. The smallest of these at Crubenmore and Loch Alvie provided the greatest opportunity for sculptural expression. This was perhaps because of their small span and the limited demands that they placed upon structural requirements. They were designed to appear as rock-like protrusions in their surroundings with the surfaces of their concrete spandrels carefully faceted by using angled triangular shuttering (*Fig. 3.6*) and whether foreseen at the time or not their surfaces have weathered so that they now blend in with the landscape (*Fig. 3.7*). The basic structure of each comprised slender reinforced concrete arches concealed behind the angular spandrels whose prime function was to contain the hardcore filling supporting the road deck. The supports for the arches and the cutwaters project well forward from each side.

This structural arrangement of arch, spandrel wall and hardcore fill was repeated in the design of the Spey Bridge, the largest of the A9 bridges, although in a far simpler visual form. The three arches of the Spey Bridge were of increasing span because the roadway climbed as it crossed the river (*Fig. 3.8*). The spandrel walls were curved throughout their height, rather than being faceted, producing something of the appearance of a dam structure. The curve also continued above the road level to form the parapet. It seems as if the outward slope

Fig. 3.5. The modified approach spans of the Findhorn Bridge.

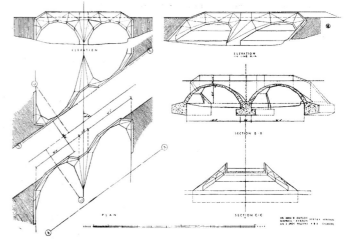

Fig. 3.6. Drawing of one of the smaller A9 bridges showing the faceted concrete.

Fig. 3.8. Drawing of the Spey Bridge.

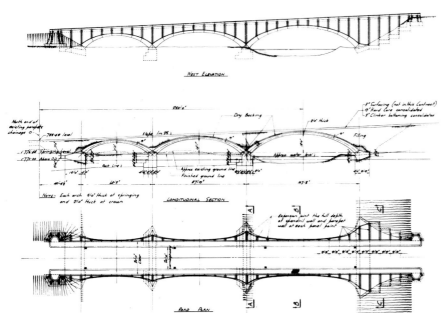

Fig. 3.7. One of the smaller A9 bridges today, weathered into the landscape.

of the spandrel walls in all these bridges was derived from the angle of repose of the hardcore fill so that the supporting arches had to be much wider at their springings than at the roadway. However, at the Spey Bridge the arch is more clearly differentiated from the spandrel walls above. Again, the careful detailing continued through to the ground with pointed projections at the arch springings and the bridge has something of the overall form and feel of a masonry structure (*Fig. 3.9*). The nature of its concrete construction was expressed in a series of wide vertical expansion joints that divide the spandrel wall and the parapet into a series of panels, each 9ft 9in. long.

Structural efficiency versus visual effect

The impression given by these schemes, produced between 1925 and 1926, is that engineering efficiency was secondary to the visual effects; that there was a certain amount of 'over-stucturing' to achieve these effects. However, this is explained in replies which both men made to a lecture given by Professor Beresford Pite in March 1925, entitled 'The aesthetics of concrete'. While there were other interesting issues arising

Chapter Three
Bridges and aesthetic theory

from this lecture, it was Pite's suggestion that reinforced concrete by its very strength produced proportions that would revolutionise architectural design, which provoked their responses. Williams took this cue to explain the main reasons for the prevailing attitude that concrete was a cheap material whose structural forms defied 'all the canons of architecture'. He claimed that the main reason for this attitude was that architects treated the structure of their buildings as completely independent of the façades, simply passing the structural design to specialist firms for competitive tender. In these circumstances it was hardly surprising if the structures appeared so 'spindly'.

'I maintain that if you put out to tender on the basis of design coupled with cost, you would be staggered at the dimensions to which engineers would reduce brick columns. Then, I suppose, architects would say "this looks very skinny, I shall have to put some more brickwork round the columns", and they would put it without a bond, and the result would be a sham.'[56]

Of course, masonry had developed without the aid of calculations so that with the sizes commonly used it was relatively lightly stressed. In contrast, reinforced concrete had developed scientifically and so could be designed to higher stresses. While this had the disadvantage that the resulting slender structures produced the attitudes described above, its major advantage was that designers could design freely with it knowing what its minimum dimensions were:

'The early history of reinforced concrete has the advantage, however, that we had to start from zero; we know what is the minimum amount of material which can be put to do the job from the point of view of cost first. After that, other considerations will come in, such as the matter of permanence, the matter of beauty. From the zero start can be built up a structure which is not a building of affectation but one which has grown; not a robust state of health by a process of evolution from zero.'[57]

We can see from this that at that time Williams did not equate structural efficiency with beauty and seemed quite at ease with the idea that the final form should be detemined as much by aesthetic requirements. Not unnaturally Ayrton, in this reply to the same lecture, endorsed this view:

'…why should we sit down and imagine that in future we have to design spiders' webs? It is not fair to the material…It is only engineers who have the courage to use the material as architecturally as they can; and it is for us now to get down to it and see what we can do, treat it seriously and generously, or we shall lose all the work which is going in that direction.'[58]

These clear statements, contemporary with the work on the early Scottish bridges, perhaps explain their over-structured and rather sculptural qualities. They also show how the two men appeared to be at one in consciously designing without regard for complete structural efficiency in an attempt to produce visually appealing structures. The assumption of both men seems to be that if brick masonry architecture was over-structured there was no reason why concrete architecture should not be also. Whether or not the lead in adopting this philosophy came from Ayrton, Williams' attitudes were to change because shortly after 1926 he began to produce his own technical and general design articles which were concerned with structural efficiency in bridge design. He then appears to have returned to a view that design should be governed more by structural efficiency based on the cost implications of design, a development which is reflected in their later structures.

Wansford Bridge
It was from 1926 that Williams began to impose on the collaboration a greater respect for the structural integrity of their designs, the turning point being their design for the Wansford Bridge on the Great North Road (A1) in Huntingdonshire.[59] An important design requirement of the Ministry of Transport was that the completed structure should harmonise with the vernacular tradition of adjacent buildings. For this reason, a three-arched structure was proposed, very similar in scale to the Spey Bridge. But while, at Spey, it was the

Fig. 3.9. Spey Bridge in use shortly after completion.

spandrel walls retaining the hardcore supporting the road that dominated the design, at Wansford a series of arches carried the road and were clearly expressed (*Fig. 3.10*). This was much more like a traditional masonry structure with parallel-sided arches. Williams also decided that as he was producing a traditional masonry-like structure he would also design it as a compression structure, eliminating the need for reinforcement. Consequently, the Wansford Bridge was designed as a mass concrete structure, using as a precedent a similar American example at Sidney, Ohio.[60] Although structurally a three-hinged arch, this is not apparent from the bridge's appearance because the thickness of the arches are not reduced at the hinges and at the springing, as strict expression of the structure would seem to require (*Fig. 3.11*).

Chapter Three
Bridges and aesthetic theory

Fig. 3.10. Wansford Bridge on completion.

Perhaps this was at the suggestion of Ayrton. Certainly the result seems much more expressive of mass concrete because of this.

The balance between the use of mass concrete and reinforced concrete was such that, although the former required more concrete, there was a corresponding saving in the costs of reinforcement. Thus, if the aggregate were readily available, so reducing transport costs, mass concrete could prove the cheaper. These were indeed the conditions at Wansford. There was a ready supply of aggregate on adjacent land and its isolated location would have increased the price of reinforcement because of transport costs.[61] These were factors that would have influenced Williams in his aim to produce an efficient structure. Curiously, the completed bridge looks less like a mass concrete than do the Spey, Crubenmore and Loch Alvie bridges. These appear to be of mass concrete because of their solid spandrel walls, while, at Wansford, the road is carried by a series of secondary, semicircular arches, thus producing open spandrels. The effect of this is that a mass concrete structure appears much lighter than those using reinforced concrete arches.

The part of the bridge that more closely resembled the Scottish bridges in its structural arrangement was in fact far less visible. *Concrete and Constructional Engineering* noted that *'The flood arch and cattle creep is designed to flare out to the natural slope of the embankment so avoiding the use of wing walls'*.[62] This was like the arrangement of some of the A9 bridges but there is no concrete to contain the fill above the concrete arch. The result is that this flood arch is hardly visible and it is not featured in most photographs of the bridge.

Open spandrel arches

In his arch bridges built in Scotland, Williams continued to use the open spandrel form that he had begun to use at Wansford but returning to reinforced concrete produced more open, lighter looking structures. At Duntocher the verticals were part of frames that carried the road deck. The journal *Concrete and Constructional Engineering* particularly noted this bridge for the way in which the arch structure was handled. Compared with other arch bridges, its appearance is rather clumsy because the arch itself is not handled in the normal way as a continuous member. Instead the bridge deck is extended downward at the centre to take compression from the arch members either side (*Fig 3.12*). *Concrete and Constructional Engineering* commented that it *'appears logical…to run the arch into a solid slab at the crown, making full use of the necessary deck slab for the purposes of the arch itself'.*[62] The bridge is straightforward almost to the point of crudeness with the ends of the transverse frames clearly visible below the deck. The now demolished Carrbridge was more elegant, with the frames less prominent and the incorporation of the bridge deck as part of the arches not so clearly articulated.[63]

The road deck of the now redundant Dalnamein bridge has a flat road deck on columns with cruciform capitals.[64] The cruciform capitals were a more visually satisfactory solution because they avoided the rather obtrusive transverse beams used at Duntocherer and Carrbridge. Moreover, these capitols are reduced to what can only be described as small triangular brackets that can only be seen from below. All these were built between 1926 and 1928, and Ayrton's decorative effects imposed then are very simple, subtle embellishments compared with the treatment of earlier bridges. In the Dalnamein Bridge the parapet steps upward toward the centre, the steps being at divisions in the bridge deck below. It also turns outward at its base helping to conceal the structure under the deck.

Fig. 3.11. Wansford Bridge under construction.

Fig. 3.12. Duntocher Bridge

43

Chapter Three
Bridges and aesthetic theory

Double cantilever bridges

Ayrton seems to have had far more influence on the double cantilever bridges at Lochy, near Fort William, and at Montrose. The dates on the tender documents show that these were being handled by Williams' office at much the same time and they show some similarities in their approach, particularly in the early design for the former. Here Ayrton began to reapply decorative motifs to their surfaces in order to create visual interest and proposed decorative features to mark the entrance to the bridge. The first proposal was for a bridge with only two pairs of double cantilevers. The structure on each side was between the roadway and the footpaths that were cantilevered outside the main structure (*Fig. 3.13*). These cantilevers required a much deeper structure over the supports which was provided by raised sections, clearly expressive of the bending moments, and to these Ayrton proposed to apply radiating lines; hardly an expression of the structure. He also proposed what looks from the drawing to be a mock medieval gatehouse at one end. In the design as built, the number of supports and spans was increased so that the spans, and hence the moments to be carried by the cantilevers, were reduced. They could now be accommodated within the parapets that were raised slightly over the supports where the moments were greatest (*Figs 3.14 and 3.15*).

In this design the circular columns were bought to the outside edge of the structure to be under this parapet (*Fig. 3.16*) and this gives the bridge a rather heavy-footed appearance. Ayrton seems to have been unwilling to let this structural arrangement stand by itself. His approach features were reduced to towers with arches between them and the radiating decoration was modified but, if anything, had less to do with the pattern of reinforcement within.

Montrose bridge, the largest and most prestigious bridge that Williams and Ayrton designed together, presents something of a puzzle because Ayrton not only appears to have been successful in rejecting Williams' desire for simplicity but seems to have persuaded Williams to adopt a strikingly unusual structural form. Built at Montrose, Scotland (1927–30), the bridge appears to be a reinforced concrete suspension bridge (*Fig. 3.17*) although one can see from this view that it is in fact a double cantilever bridge with a suspended central span. In fact, the model that was made of the bridge had the central span removable to demonstrate this. The hangars supporting the side span decks are heavily reinforced concrete (*Fig. 3.18*). The intention here may have been to give the road user the arcade effect that they had successfully produced at Findhorn or to produce some historical reference to the 100-year old suspension bridge that it replaced. Whatever the reason, *Architect and Building*

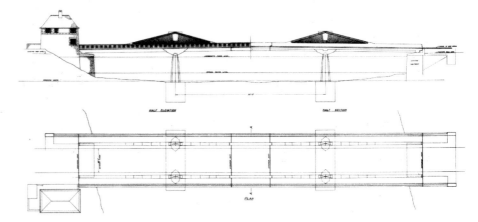

Fig. 3.13. Original design for Lochy Bridge.

News, ever ready to praise Sir Owen, was not about to criticise this design but contented itself with rather general praise and rather ambiguous comment.

'*A new Owen Williams Bridge is always something of an event. This is because each new bridge built by Sir Owen shows new thought in the aesthetic treatment of constructional form, particularly when the material used is reinforced concrete. As progressive efforts to formulate an expression for this new material these bridges are of greatest interest.*'[65]

However, an article in *Engineering News Record*, which described and illustrated the bridge, drew some criticism from American engineers and was followed by an exchange of correspondence in which Williams defended the scheme on economic and structural grounds.[66] Williams' arguments do not appear particularly convincing. He claimed, in contradiction to the statements made by the American correspondents, that it was cheaper than an equivalent steel bridge but he failed to produce the detailed figures that one of the Americans demanded as proof of this. Without any figures we are unable to judge.

Whatever the logic of the overall form, the details are handled with care, clearly expressing the structure. For all the initial appearance of a suspension bridge, it is clearly a double cantilever rather than a true suspension bridge so that the parapets form

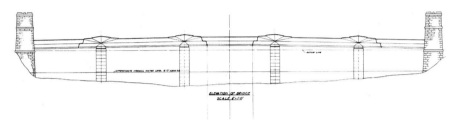

Fig. 3.14. Final design for Lochy Bridge.

Fig. 3.15. Lochy Bridge — reinforcing drawing showing raised parapet and reinforcing for cantilever action.

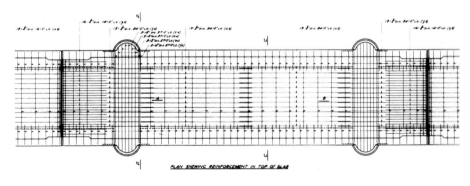

Fig. 3.16. Reinforcing plan of one section of the Lochy Bridge.

Chapter Three
Bridges and aesthetic theory

the compression chords, expressed in the solid junction between these and the top chords. The pylons and hangars are lightened in appearance by their changing sections and the central suspended span is clearly differentiated from the structure of the cantilevers. The other carefully controlled details are the hinged supports. As at Findhorn, this bridge successfully uses structural expression to provide both the views of and from the bridge.

Bridge theory

Williams had produced a forceful argument in favour of the use of mass concrete arches in his response to Pite's lecture. In view of this successful demonstration at Wansford, it is appropriate to quote his remarks at length.

'I find that [Professor Pite] has discarded arches; he feels that arches are no more. I have a very definite opinion that the only permanent structural element is the arch. It is the only member in which tension can be eliminated, leaving only compression; that is to say, that the structure then depends entirely on the force of gravity...The introduction of steel into concrete pre-supposes tensile stresses, the introduction of steel also means corrosion. Professor Pite imagines the ancients using reinforced concrete for monumental buildings; I feel they would never have used reinforced concrete for monumental buildings. The function of reinforced concrete is as a commercial expedient for the production of cheap buildings to last a period not exceeding 100 years...Had the ancients known the secrets of concrete as we know them, they would have built in concrete without reinforcement, and that they would still have built in arches...'[67]

Reading this today, one wonders whether Williams was always so outspoken or whether conventions were different at the time; Williams was giving the vote of thanks for the lecture. While there is an argument here about temporary versus permanent structures, the real significance of his statement is in his advocacy for mass concrete as a means of achieving monumental structures. The sequence of events is perhaps significant here. The article describing the American mass concrete bridge appeared in 1923, his outspoken response to Professor Pite was two years later when he was working on Wansford bridge in which he was relying upon the very principles that Pite was rejecting.

At that time Williams was developing his ideas on the structure of bridges with two complementary papers. The first was delivered to the Institution of County and Municipal Engineers and was concerned with the design of beam and slab concrete highway bridges (1926). In his introduction, Williams suggested that as reinforced concrete was such a flexible material, cost was the necessary starting point in deciding on the structural form.

'The very flexibility of reinforced concrete as a designing medium makes it difficult to settle such questions (of structural form) without a general investigation into cost. There are no standard sections of beams or troughing to guide the designer; he is free to make sections as he pleases and at any spacing. Indeed, it is at once a delight and a difficulty of the material that it is so untrammelled with conventional dimensions.'[68]

The paper then examined the cost implications of various combinations of slabs, with or without beams, at a variety of centres. In doing this Williams was able to produce a table broadly outlining the economic range of various alternatives, enabling engineers to select an appropriate arrangement for a variety of spans.

The second paper, entitled 'The philosophy of masonry arches' (1927), applied modern structural analysis to the ancient arch forms and examined their relevance to modern conditions. Williams maintained that these modern conditions had radically changed the design of bridges because modern traffic demanded a flattening of the bridge structure, ever increasing spans and a need to accommodate heavier loads. By analysing 200 arch structures of spans up to 300 ft, Williams was able to present criteria that would enable designers to assess the structural limits of tension-free structures and to show when reinforcement would be required.[69]

However, if he had hoped that this would demonstrate the general applicability of mass concrete arches, he was to be disappointed. The increased volume of concrete required only made it applicable for smaller spans and where aggregate could be obtained cheaply.

In developing these papers Williams was tackling the problem facing any designer, that of where to begin and what overall proportions to use for the structure. He was also becoming more closely committed to the idea of simplicity in the design of concrete structures. He insisted that beams were an unnecessary extravagance until a span of at least 20 ft was reached, his main argument being that the increased simplicity of the shuttering greatly reduced the cost. Neither were his recommendations for simplicity in concrete design based entirely on its economic merits. In an article directed to a wider audience, he made a call for simplicity as a means of achieving concrete forms of architecture more closely related to the material's properties:

'*A simple, direct, even unmathematical*

Fig. 3.18. Montrose Bridge under construction.

Fig. 3.17. Montrose Bridge.

Chapter Three
Bridges and aesthetic theory

outlook on concrete and reinforced concrete will alone shape the material into its own peculiar forms, instead of imitating the forms which other materials have developed for themselves…

The growth of concrete and reinforced concrete in the next 21 years would be phenomenal if all engaged in the industry took as their guiding principle — Simplicity.[70]

This helps to explain the development of his bridge designs, which began to rely upon structural efficiency and simplicity for their aesthetic appeal. With Williams turning his mind towards theoretical issues and addressing efficiency in bridge designs, the later bridges need to be considered in this light. While the early bridges have clear and distinctive architectural intentions behind their form this does not appear to be true of most of the later bridges whose form seems to come primarily from engineering considerations.

This is not the place to discuss the aesthetics of bridges in any detail but their design does present some quite distinct problems. Designing for the experience of the bridge user, the view from the bridge, can be quite different from designing for the experience of the external spectator, the view of the bridge. Of course, the relative importance of these two will depend upon its location. Ayrton made it clear that he was concerned with the former in his treatment of the Lea Valley Viaduct and he seems to have carried this concern through to his much more successful handling of the Lochy Bridge. What he provided for the former was in much the same category as placing statues of lions on pedestals at the entrances. The difference was that his 'lions' were larger than normal and in the modern abstract manner. Using the structure of the bridge to provide the aesthetic experience as at Findhorn was much more satisfactory and the view of the bridge was as successful as the view from it. Readers may be aware of other bridge forms that provide this kind of effect, although this bridge may well be one of the better examples. In contrast, the Spey, Crubenmore and Loch Alvie bridges were clearly conceived with the view of the bridges in mind. Possibly simple beam structures would have been more economical for the short spans involved in the last two of these.

The other issue is the extent to which the structural form of the bridge should determine its aesthetic qualities. When dealing with spans that require an arch — and over deep and fairly narrow valleys little else will do — the basic form is given and, as we can see from the various designs produced, it is a matter of handling the relationship between the deck and the supports. It would hardly be possible to make any visual statement of either approach or crossing without detracting from the overall form. The effect of the bridge is determined by the structural arrangement and the move towards open spandrel structures placed the initial determination of form clearly in the hands of Owen Williams. What is not clear is the reason for the change in arrangement between the rather clumsy looking Duntocher Bridge and the more elegant Dalnamein Bridge.

Did Williams believe that structural expression alone would produce a satisfying result for the simple beam on columns arrangement of the Lochy Bridge? This would seem to account for the expression of bending moments in the form of the parapet adopted both in the original design drawings and in the final form. If he did think this Ayrton certainly did not and in the original drawings produced a decorative treatment that separated the horizontal balustrade from the upward projections. With all the structure of an arch bridge below the deck any such conflict between structural expression and architectural treatment of the view from the bridge is avoided.

Ayrton's contribution
While Williams may have optimistically supposed that Ayrton shared his intentions, in retrospect we can see that their objectives were merely similar rather than identical because while Williams was interested in a concrete architecture Ayrton was looking for

a concrete style of architecture. At the same time, Williams, ever the intellectual, was developing his ideas about the structural form of bridges, comparing the economics of the different forms available as we have seen from the papers that he produced. The practical results of this can then be seen in the later inter-war bridge designs in which he moved away from the rather novel visual forms that characterised the early collaborative examples to much simpler and, to a large extent, more conventional designs.

As noted at the beginning of this chapter, it is difficult to disentangle the contributions of each member of a collaborative design and this is particularly difficult where it is clear from events that one of the partners is developing his approach. Perhaps if Ayrton had not been involved in the later 1920s bridges that they built together, Williams might have realised earlier his aim of structural simplicity. The evidence (to be considered in the next chapter) is rather that Ayrton remained rather conservative in his approach, evident in the increase in applied decoration to the structurally simpler flat-decked bridge schemes. Williams eventually produced his simplest structure in the speculative design that he did alone for Waterloo Bridge, a design that was admired by the aficionados of the Modern Movement but rejected by the establishment.

Waterloo Bridge

The design for Waterloo Bridge was produced in March 1932, prepared secretly within the office and delivered to the LCC when complete. The unsolicited proposal was presented to the LCC together with a firm commitment to undertake the entire project, including the demolition of the old bridge, for a price of £693 000 in three and a half years.[71] Its structural concept had a bold self-asserting simplicity devoid of any decorative treatment. A single row of 20 ft diameter columns under the centre of the deck at 150 ft centres were to be used to support cantilevered slabs, projecting 40 ft in all four directions. Between these, suspended reinforced concrete boxed-decks, 10 ft deep, were to be cast (*Figs 3.19 and 3.20*).

Although this design received extensive publicity in the press, for such an important location its simple visual appearance was much too far ahead of its time for it to have received the approval of the influential institutions that could have made it a reality. One of the most frequently quoted criticisms was Herbert Morrison's remark that it was 'a mere roadway on upturned drain pipes utterly unworthy of the site'.[72] The general consensus was that for an important Thames bridge a traditional arch structure was the only visually appropriate solution. The director of the British Steelwork Association, C. J. Kavanagh,

Chapter Three
Bridges and aesthetic theory

seized upon this general criticism of the proposal to promote his own natural preference for a steel structure which he presented in a letter to *The Times*. Williams then responded to this through the pages of *Concrete and Constructional Engineering* quoting Kavanagh's letter, where he said:

'There would not be the slightest difficulty in constructing more economically a steel bridge of attractive appearance of the arch type similar to the other bridges over the Thames. As an illustration, a bridge recently built in Cologne is considered by many to be the most beautiful in Europe. It is built entirely of steel, it is nearly one and half times as long as Waterloo Bridge, carries six lanes of traffic, provides an uninterrupted waterway for river traffic, and cost less than £600,000.'[73]

Williams' reply exposes Kavanagh's fallacious comparisons and notes his failure to include the cost of demolition work in his quotation. Referring to the issue of concrete versus steel, he wrote:

'Not being interested in propaganda of a particular material, I naturally investigated stone and steel as well as reinforced concrete before making my offer, and can speak with confidence on the relative economies.'[74]

Given Williams' obvious predilection for concrete one might wonder whether he was not being a little disingenuous here. His drawings show that he did produce a steel proposal, although it adopted the same

Fig. 3.19. Waterloo Bridge showing the box section that formed its structure.

simple form as his concrete scheme with the main support structure in concrete and the intermediate decks in structural steelwork. Perhaps an arch might have been more appropriate for steel. Despite Kavanagh's attempt to promote the vested interests of the steel industry, neither proposal was adopted. Instead Sir Giles Gilbert Scott was appointed to produce a much more conventional scheme, which was designed with a masonry arch veneer to clad his engineer's steel and reinforced concrete structure.[75]

While Williams was disappointed that his scheme was not accepted, the publicity it provoked was helpful in exposing his views. It was also helpful to the small group of modernists in architecture who used the controversy to illustrate their own ideas for producing an architecture that truthfully responded to the properties of modern materials and modern conditions. P. Morton Shand, for example, was so

Fig. 3.20. Waterloo Bridge — design perspective.

impressed with Williams' scheme that he sent the drawings to Maillart for comment. Maillart replied that he liked its elegance but questioned some of the engineering details. 'How interesting', Shand claimed, 'that misinformed British architects and public opinion should criticise Williams aesthetics while applauding its engineering'. Explaining this he wrote:

'…our eyes have grown accustomed to arched spans because brick and stone can only be built in that manner. When, therefore, we are brought face to face with another material that is able to ignore the arch of convention we upbraid the author's choice of material, his lack of 'taste', or his inexcusable disregard for the aesthetic sensibilities of others. This, if we only knew it, is equivalent to reminiscence over resource, approximate over precise, amateur capacity over technical proficiency and waste of space and material over exact calculation of how much of each is required.'[76]

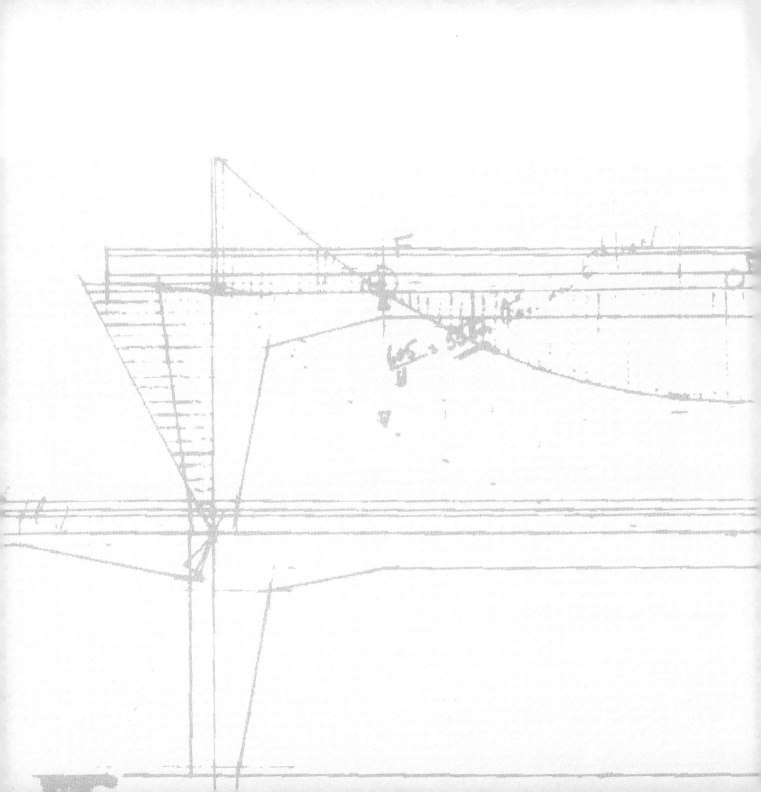

Chapter Four
The break

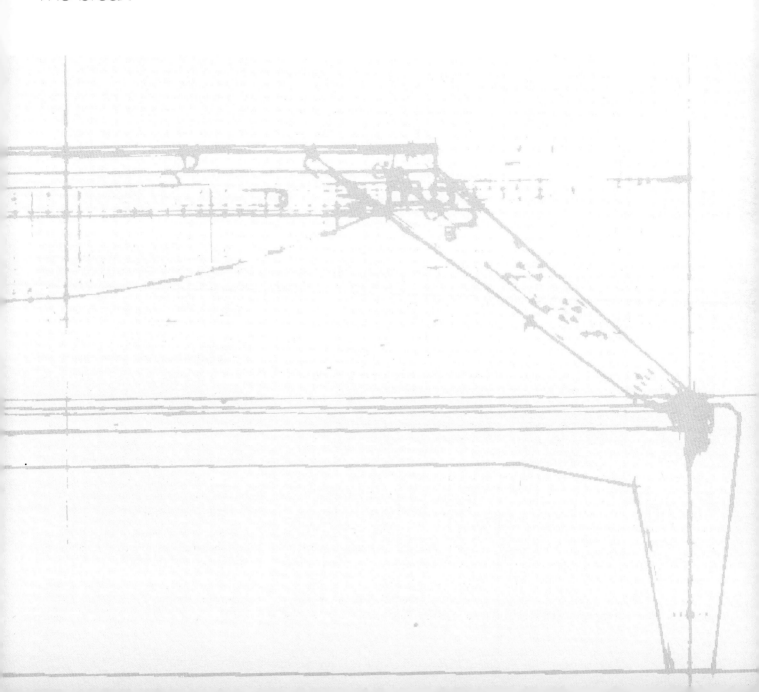

Chapter Four
The break

During the time that he was collaborating with Ayrton on the bridges, Williams was not only developing his ideas about bridge design but was also developing a wider interest in architectural design through his membership of the Architecture Club. He would have been aware of the concrete architecture being produced on the continent but he was also introduced to the work and thories of the German architect, Erich Mendelsohn through a book on his work that was given to him in 1926.[77] This book contained sketches by Mendelsohn as well as examples of his completed works. The latter, which included some industrial projects, were described in some detail with photographs and drawings. The industrial projects were a hat factory at Lukenwalde – which was largely a collection of sheds based on reinforced concrete frames but with a dramatic roof over the fabrication and dye works, presumably to provide ventilation – the building for the Berliner Tageblatt and a power station for the Myer-Kauffmann textile works at Wüstegiersdorf. There can be no suggestion that Williams was influenced by the visual form of these buildings because Mendelsohn's handling of concrete surfaces was rather heavier than anything he was to produce. What would have influenced him was the rather short text in the book that comprised extracts from his lectures. These express a clear functional ideology.

There were short extracts from two lectures, one entitled 'The problem of a new architecture', given in 1919, which is partly concerned with the logic of construction materials. But more significant is his 1923 lecture on 'Dynamics and function' in which, speaking of the forces in a steel structure, Mendelsohn says:

'It is our task to find an architectonic expression for these forces of mobility, and by means of architectonic form to establish an equipoise for these tensions as well as to master the inner forces which are bent upon expressing themselves in outer forms.'

And of functional planning he says:

'We architects must from the very beginning subject our plans to the demands and limits placed on them by material and constructional considerations. We must regard them as the preliminary conditions upon which the organization of an entire building depends.' [77]

While there is little in common between the romantic expressionism of Mendelsohn's work and Williams' 1930 projects, these clear statements of his design approach seem to have confirmed Williams in his own thoughts on the engineering basis of modern architectural design. It would have helped to shape his attitude to bridge design. Given the theoretical work that he had done on bridges that formed the basis for the papers that he gave, combined with this approach to aesthetics, one can imagine that he would have been interested not only in simplifying the form of his bridges but also expressing their structures. Thus, we can see the form of his Waterloo Bridge proposal as a refinement of the simple forms used for some of his realised bridge designs. The kind of architectural embellishments that Ayrton had been adding would hardly have fitted in with these Mendelsohnian ideas. Williams and Ayrton seem to have been moving apart in their ideas. But the acid test of the effectiveness of their collaboration was surely when Williams and Ayrton returned to building design.

One might infer from their earlier statements that their intention after Wembley was to work towards the production of concrete buildings the forms of which were directly related to the properties and techniques of reinforced concrete construction. But perhaps this is to assume more from Ayrton than he intended. He was certainly concerned with the surface treatment of concrete and modified his views on that in the course of their collaboration. When the *Architects Journal* produced its special issue on concrete in 1926 Ayrton contributed an article on concrete surfaces. He said that:

'The strong prejudice against concrete has grown entirely from neglect or lack of vision in seeing in it a material with aesthetic possibilities.

If the cement face, which is in itself of no practical value to the concrete, is removed and the aggregate from which the concrete is made is exposed, at once a variety of texture and colour is revealed which has an individuality and attractiveness all its own.'[78]

Nevertheless, this does not seem to have resulted in any enthusiasm for the use of the material in his buildings.

His designs for Bedford College, London University are in a restrained neo-classical brick style that was common at the time, as was his designs for the National Institute for Medical Research, at Frognal, Hampstead (1929–30).[79] The building that must most clearly have suggested to Williams the futility of trying to work with architects was a warehouse built for the Pilkington Glass company in Hoxton, North London (1928–30).[80] The architect credited with this is Sir John Simpson, for whom Ayrton had been working, and although by this time Ayrton had his own practice it seems likely that he was the author of this particular building. It certainly resembles the Frognal building in its treatment of the brickwork. Even though the buildings had entirely different functions, certain features of their façades were identical. For example, both make use of semi-circular window heads, with windows extending two to three floors above ground level. As an industrial building one might have expected Pilkington's warehouse to show some expression of the

Fig. 4.1. Wrigley factory by Wallis Gilbert and Partners.

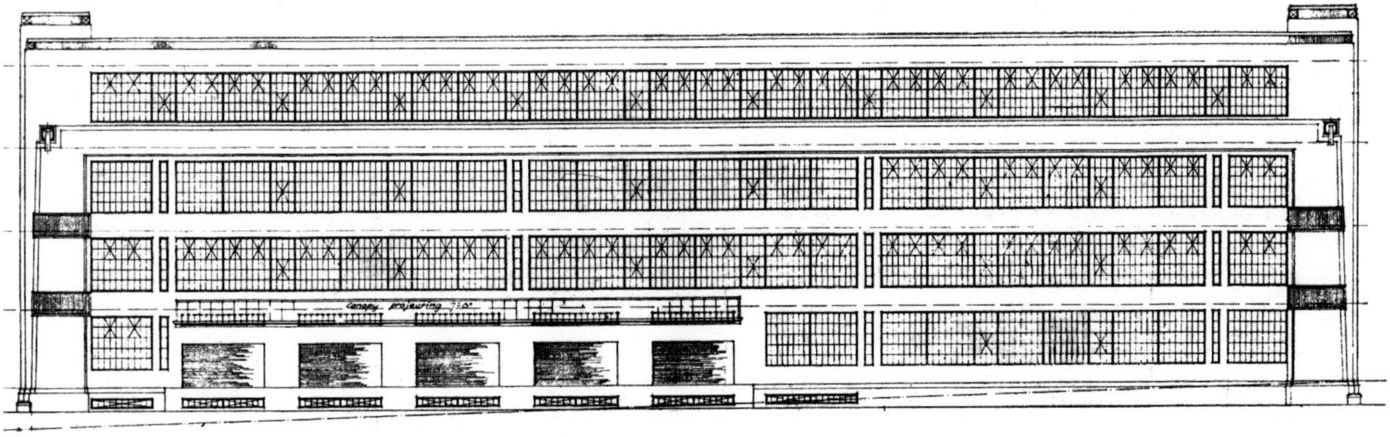

Chapter Four
The break

concrete structure but Williams' post and beam frame concrete structure was only notionally respected in the elevations, with the brickwork piers arranged to coincide with the concrete columns behind.[81]

Ayrton seems to have absorbed nothing of Williams' ideal of simple structural forms during their collaboration and in this factory design Williams was simply producing a structure for a quite different architecture; again he found himself producing buildings whose architectural features owed nothing to their reinforced concrete structure. Given the function of the building one might even have expected to see flat slab construction used so that the fact that it had a conventional frame suggests that it might even have been planned without any input from Williams.

By this time, with Williams developing his ideas about architecture he must have been aware, through Bennet and Yerbury's book, of the kind of concrete architecture that was being put up elsewhere in Europe.[82] He would also have been aware of the factory buildings being designed by Thomas Wallis in collaboration with his old firm Trussed Concrete Engineering that made frank expression of their concrete structures. Wallis' use of flat-slab floors and set back columns enabled him to give a strong horizontal expression to his buildings (Fig. 4.1); a new effect that had been welcomed by *Architect and Building News*.[83] Given these developments, might not Williams have hoped to be involved in something a little more forward looking, especially in an industrial building? Was this not a movement in architecture in which he might have seen himself playing a part? Instead, he found himself collaborating with someone whose architecture might not unkindly be called reactionary. Ayrton's inability to escape the architect's traditional stylistic role was effectively denying Williams the opportunity to exercise his own creative skills and others could be seen to be exploring the ground that Williams might have regarded as properly his.

There can be little doubt that it was the complete failure of Ayrton to approach his objective of a modern form of concrete architecture, as well as the fact that he was, in effect, reverting to the role of technical assistant in the design of buildings, that provoked Williams to abandon the idea of collaborating with architects. What must have been even more galling was that shortly after the completion of these buildings Ayrton lectured at the RIBA on the aesthetic treatment of modern bridges (April 1931); something that he was only qualified to do through his collaboration with the man whose developing ideas he had failed to keep up with. Williams made his position clear in the comments that he made following this lecture. Williams asserted that after eight years of collaboration with Ayrton, and despite 'many difficulties', they were still friends but he then offered the opinion that the philosophical standpoints of the architectural and engineering professions were irreconcilable. He had come to believe that architects and engineers were approaching design from quite different directions and that to begin with the idea of creating a work of beauty was absurd. He believed that beauty came first from being honest: *'Be honest first and if you are honest you will be beautiful, but do not attempt to be beautiful and dishonest'.*[84] Here he seems to have been echoing the ideas of Mendelsohn.

As for the possibility of collaboration between the two he then used a very telling analogy:

'I was walking through the Park today and I saw a very beautiful girl…With her were two young men. And as they walked by, both happened to be on the same side of this young lady. And it occurred to me, as a matter of observation…that the man on the outside was in a quandary, and I thought, would it not be better if he were on the other side of the young lady? But I think he must have been of the opinion that he had lost it, but that, keeping on that side of her, he would get a few stray glances he would not get if he were on the opposite side to his competitor. And I think that those two competitors for beauty represent very much the position of effect and

practicality. I do not know which should retire, but I do think one of the two gentlemen ought to retire in favour of his opponent.' [84]

He was clearly referring to his relationship with Ayrton. We may suppose that that beautiful girl represents the architectural commisions while he and Ayrton are the two young men. But which is which? The authors leave it up to the readers to choose for themselves, always assuming that Williams had a clear idea himself.

This is a complete reversion of his earlier position in 'Concrete as a Partnership of Engineering and Architecture' (1924). Although he maintained his view that architecture and engineering of themselves were essentially identical,[85] his experience with Ayrton and other architects confirmed for him that the diversion of the two professions over a long period of time had resulted in two groups with polarized views on the nature of design. Therefore, he concluded that, given the very different positions they had adopted by the 1930s, partnership between them was impossible.

The opportunities

It was in the years 1929 and 1930 that Williams had his real chance to practise as an architect as well as an engineer. During this period he obtained commissions for three major buildings, the pharmaceutical factory for Boots at Beeston near Nottingham, and two buildings in London, the Dorchester Hotel in Park Lane and the Daily Express building in Fleet Street. Williams has never been fully credited with either London building although for quite different reasons. They both came to him by way of architects previously engaged and their acquisition by Williams requires some explanation. But at the time it was the commission for the Dorchester Hotel that must have pleased him most because, as Richards observed:

'*The commission meant a lot to Owen Williams because he had always wanted to be not only an engineer but an architect…A factory in the Midlands is one thing and a luxury hotel on the fashionable centre of London quite another.*'[86]

Thus, Williams' opportunity to operate as an architect came a little before his comments at Ayrton's lecture (see above); the comment that he made was simply a public announcement of a change of position that had already taken place.

It seems that although the Dorchester and Daily Express need to be considered first in terms of their 'completion' by the office, the Boots factory was the first of the commissions. The job number of the Boots building (no. 378) predates that of the Daily Express and Dorchester and shows that design work on this building started early in 1929 (although drawings now in the possession of the Boots Company are dated

Chapter Four
The break

1930 and 1931). There was nothing particularly remarkable in his acquiring the Boots commission. By this time Williams had a successful practice as a consulting engineer and it was not uncommon for engineers to be designing factory buildings without the involvement of an architect. The site for the building was a 300 acre piece of open land bought by the United Drugs Company in 1926. They had already put up a small soap factory on this site in 1927 but the firm responsible[87] were not interested in designing the enormous complex that the company now had in mind and so Owen Williams was approached.

His remarkable appointment as architect to the proposed Dorchester Hotel in Park Lane, London, in November 1929 needs a little more explanation. The Dorchester House Syndicate Ltd, the corporate client for the proposed hotel and part of Gordon Hotels, clearly intended their new hotel to be a 'modern' building from the outset because in 1928 the first architects commissioned were Wallis Gilbert and Partners.[88] The firm was just then completing work on the late lamented Firestone Factory. Thomas Wallis, the active partner, was best known as a factory designer in the modern manner with an ability to handle concrete and his buildings for the Wrigley Company at Wembley had been completed in 1926. That and his building for the Gramophone Company at Hayes in Middlesex had both attracted favourable comment in the architectural press and no doubt a similarly modern form was expected by the proprietors of the Dorchester Hotel.

Little progress was being made on the design because of a protracted debate with the consulting architect to Gordon Hotels, P. Morley Horder, whose main brief was to ensure that the completed building harmonised with the surroundings, as envisaged by the Duke of Westminster's consultant Sir Edwin Lutyens.[89] When Wallis Gilbert and Partners failed to produce a viable scheme within the client's time limits, to make up for lost time the latter decided that they needed a different architect. Sir Robert McAlpine and Sons were the largest shareholder of the Dorchester House Syndicate Ltd and Sir Robert McAlpine himself was one of the few directors. Williams' architectural ambitions were well known to McAlpine; they had become close friends through their joint work on many buildings and bridge contracts over the preceding decade, the first of which had of course been the Empire Exhibition Buildings.[90] Thus, it was that following the failure of Wallis Gilbert to produce a scheme, McAlpine approached Williams to take over the project. Williams was presented with an outstanding opportunity to prove that his functionalist design theory could be applied not just to a modern building design but to a prestige building.

Not unnaturally his appointment aroused great interest and speculation in the press, although there was surprisingly little comment in architectural publications. Headlines such as 'Engineer instead of architect'[91] and 'Utility in new buildings – the engineer and the architect – who will be master?'[92] appeared in the daily newspapers, with correspondents claiming that the appointment of Williams for the design of a prestigious hotel building represented a direct challenge to the architectural profession and to traditional building forms.

His appointment to the Daily Express for their building in Fleet Street was far less dramatic and did not cause the same stir because the original architects were retained. H. O. Ellis and Clarke, a firm who were currently designing a number of newspaper buildings for other clients, were originally appointed for the Daily Express building, presumably because of their experience with this kind of work. They were, for example, responsible for the design of a series of buildings for Associated Newspapers (all called 'Northcliffe House') which were built in different parts of the country between 1927 and 1933,[93] and their initial scheme for the Daily Express was very similar to their other buildings.[94] It comprised a

Fig. 4.2. Ellis and Clarke's original scheme for the Daily Express building.

structural steel frame clad in Portland Stone and styled in the stripped classical idiom (*Fig. 4.2*). However, the Fleet Street site was very restricted with a frontage of only 80 ft and a depth of 115 ft where it adjoined a small existing steel-framed building, built for the newspaper only a few years before. Having acquired the newspaper, Beaverbrook then acquired the site between the building in Shoe Lane and Fleet Street, enabling him to build the extension that was now proposed. In the original building, the presses had been laid out at right angles to Shoe Lane. It was now proposed to have presses parallel to Shoe Lane, but the existing steel frame, and that proposed for the new building, was based on a 25 ft by 30 ft grid that severely restricted the usable floor space and made efficient planning of the printing presses at basement level particularly difficult.

One of Beaverbrook's associates discussed this problem with Owen Williams, sometime between October and November 1929. He immediately suggested how the steel structure of the basement could be replaced by a long-spanning reinforced concrete structure, which would solve this planning problem. Seizing his opportunity, he produced an outline solution that he presented to the Daily Express Building Company the following day. They were impressed with this structural solution, with its consequent saving in floor space, which improved the planning of the press-machine runs, and immediately commissioned him to redesign the entire structure while retaining the services of Ellis and Clarke as architects.[95] Nevertheless, the relationship set up was not that of architect and consulting engineer. Williams' account correspondence shows that his fees were paid directly by the client and not by the architect, as would have been normal at that time if he had been working as the architect's consulting engineer. Moreover, that Ellis and Clarke were not retained for the Manchester and Glasgow buildings that the paper built subsequently suggests that Williams played a major role in the Fleet Street building.

The effect of these commissions was that at the beginning of the 1930s Williams had three substantial building projects in his office providing him with an opportunity to show what he could do as an architect. In addition to these he also had some smaller commissions, a small factory and warehouse for Sainsburys, a laboratory for Tunnel Cement and a garage and car park in central London. While these buildings will also be discussed, like the Boots factory, there was nothing particularly unusual in an engineer being given jobs of this kind.

Chapter Five
Dorchester Hotel and Daily Express

Chapter Five
Dorchester Hotel and Daily Express

The Dorchester Hotel and Daily Express building should have marked Williams' triumphant début as an architect. They were both major London buildings and the designs for both had a strikingly modern appearance. The Dorchester, in particular, would have established Williams as someone who had to be taken seriously as an architect. In the event, it became a serious disappointment to him and perhaps a loss to the Modern Movement in Britain.

Dorchester Hotel
When Williams acquired the Dorchester commission, the *Daily Telegraph* wrote that:

'There is much curiosity to see how the innovation will be received by the RIBA and among the profession generally. The appointment even of so eminent an engineer as Sir Owen Williams will almost certainly be regarded as a challenge to the traditional status of the architect.'[96]

It is difficult to be sure of people's feelings but perhaps events indicate that there was some collective jealousy over this appointment at the time, especially as some of the newspapers seemed to regard Williams' appointment as a direct challenge to the architectural profession. However, as if to allay such fears, Frederic Towndrow, who was then architectural correspondent to The *Observer* claimed that Williams' adoption did not represent such a challenge but was instead indicative of the changes that architecture was undergoing at the time.[97] He seemed to welcome this appointment pointing out that, historically, there was nothing unique in engineers doing architectural work, it was simply that there had been an unfortunate separation between the two disciplines that had resulted in separate professions. While he accepted that it would be dangerous if engineers were to take over the architect's role completely, in the case of Owen Williams there was nothing for architects or the public to fear. Williams was not just an engineer but an artist as well, his functionalist approach was something to which architects had paid lip service for far too long. His role, therefore, would be to point architecture in the right direction and suggest changes in architectural education that would allow architects to follow his lead.[98]

By this time Williams had already begun work on the project and it is clear from Towndrow's remarks that he had seen some of the drawings and so presumably had discussed the design with Williams. Williams had acquired a young architectural assistant, J. M. Richards, to join his all-engineer staff to help with the production drawings. Richards was at the time working in the office of Oliver Bernard who specialised in

Fig. 5.1. Model of Owen Williams' design for the Dorchester Hotel.

interior designs and who Williams had come to know through his work at the Empire Exhibition where Bernard was in charge of the displays.[99] Work proceeded rapidly and as early as 1930 construction started on site.

Unfortunately, Morely Horder (consulting architect to Gordon's Hotels Ltd) had not been approached during the design process and when eventually presented with what he saw as a reinforced concrete building he decided that it was totally unsuited to the site. As the design was by that time a *fait accompli*, he resigned in protest and in a statement to *The Observer* made his reasons clear:

'From the day of the appointment of Sir Owen Williams I was not consulted in any way as to the design, and when the final plans were put before me there seemed no alternative but to resign, as they were so complete in their manner and expression of the material that there was no hope of changing the character of the design.

I should like to make it clear that the plans and elevations put before me by Sir Owen Williams are an extremely able expression of concrete forms, and my only objection is to the introduction of this manner of building into the neighbourhood of buildings distinctly foreign in character.

The proposed building will, no doubt, have a certain freshness and gaiety when first erected, but the London atmosphere will soon make the surface very dismal and depressing. I cannot suppose that it is the intention to paint the

Fig. 5.2. Perspective of Owen Williams' design for the Dorchester Hotel.

surface every 3 years, which would be the only way to keep it at all cheerful in appearance. I question if any gain in the apparent rapidity of erection will compensate for all the difficulties in making so thin a structure architecturally satisfying from within, or comfortable to live in so variable a climate.'[100]

What is curious about this statement is that Towndrow already knew that it was not to be of plain concrete and had made a point of mentioning this in his *Observer* article not three months earlier. It seems almost as if he anticipated some objections and was anxious to allay any fears that the public might have when he wrote that:

'…we need not be concerned that the building is to be of reinforced concrete, for I have

Chapter Five
Dorchester Hotel and Daily Express

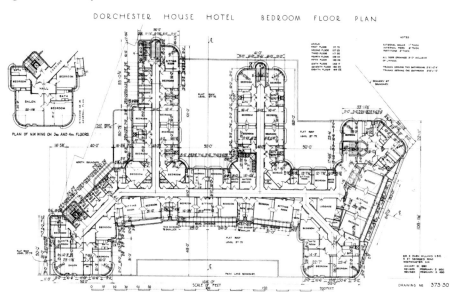

Fig. 5.3. Williams's plan for the Dorchester Hotel.

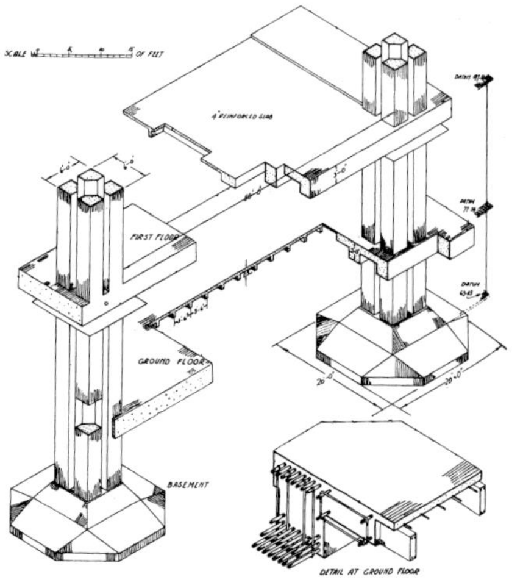

Fig. 5.4. Structural arrangement of the Dorchester Hotel (from *Architect and Building News*).

seen the material to be used and it compares with a polished stone of great beauty.'[101]

Presumably the client would also have been aware of the nature of the surface finish. Why then should Horder have assumed that it would be unpainted concrete and also apparently that it had little or no insulation? His words were an echo of an editorial in the *Architects Journal* of only a month before which, under the heading of 'Architect or Engineer?' and referring to Towndrow's praise of the building, said that:

'The obvious lacuna in the engineers' artistic equipment is due to the fact that he is concerned with individual structures only, and his scientific training does not bring him to a consideration of those subtle problems which concern the mutual harmony of several units of building which are set in juxtaposition. Let not the architect then, in a spirit of too quiet complacency, make public surrender to the engineer.'[102]

No doubt Horder would have been rather miffed at being consulted so late in the design process and may have been disposed towards making some objection but it is equally possible that Williams had failed to present his ideas sufficiently clearly. Whatever the reason, his public statement hardly left room for a compromise or reconciliation.

Williams might have been better advised to have hired his friend Oliver Bernard to work on the interiors of the hotel because some weeks later, as Williams' true intentions to avoid any kind of decoration on either the façades or internal spaces became clear, the client became hesitant, particularly at the prospect of the ballroom being conceived as a 'great whitewashed barn'.[103] Williams was asked if he would agree to work alongside Curtis Green, a well-respected architect, who would add the necessary embellishments to the elevations and the interiors. Having

64

decided to practice as an architect because of the difficulties that he had experienced with Ayrton it is hardly surprising that Williams refused to work with yet another architect and resigned, whereupon the contract was given to Curtis Green.[104] Williams' disappointment at this turn of events may well be imagined. He sent the drawings to Robert McAlpine in a taxi and the rumour is that he did not speak to him for many years afterwards, even refusing to allow McAlpines to tender for any of his subsequent jobs.[105] Here had been a first-rate opportunity to demonstrate the practicality of his theories and to make his mark as an architect on one of the most prominent sites in the West End. Instead, it was snatched away from him before it had a chance to be built but when the design was so well advanced that he must have been able to envision it clearly.

With the structure completed to the ground floor level, it was clearly impossible for the basic plan to be changed nor, with time pressing, would it have been possible to make extensive changes to the superstructure. Therefore, Curtis Green and his consulting engineers, Considère and Partners, had to accept Williams' basic design for both. While Curtis Green might have preferred to make more extensive changes all that was possible was to modify the offending elevations and to develop the interiors. Although none of Williams' drawings has survived we know

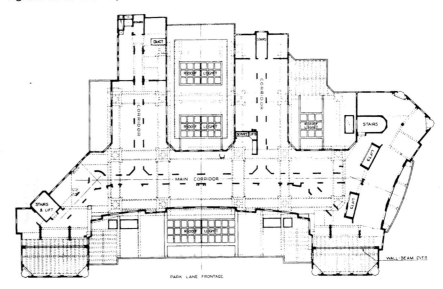

Fig. 5.5. Curtis Green's plan for the Dorchester Hotel — compare with Fig. 5.3.

what he intended for the elevations from photographs of a model that was prepared (*Fig. 5.1*) together with a perspective sketch by Keith Murray (*Fig. 5.2*).[106] Comparing these with the hotel as built, we can see that Green and his engineers made little change to the Owen Williams design.

The overall plan of the building, as devised by Williams (*Fig. 5.3*), took the form of a broad shallow U shape facing Park Lane with three wings projecting at the rear in an E shape. These wings had to be of different lengths because of the shape of the site. Within this plan the rooms were arranged in

Chapter Five
Dorchester Hotel and Daily Express

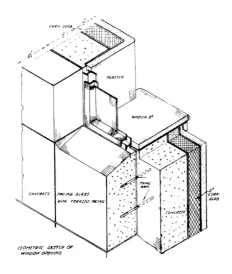

Fig. 5.6. The construction of the wall of the Dorchester Hotel with precast facing slabs (from *Architect and Building News*).

the conventional manner, either side of a central corridor. The principal structural problem in an hotel is of course the transition between the large public spaces on the ground floor and the small bedroom spaces above. This was handled by a complete separation of the two structures (*Fig. 5.4*).[107] For the large spans of the basement, ground and mezzanine floors, Williams devised a concrete frame system with large columns, 6 ft square and each comprising four separate legs. This allowed services to be run in the space between the legs. These groups of columns supported twin beams spanning distances of up to 56 ft, cast integrally with Tee beam floor slabs. The floor, immediately below the bedroom blocks was designed as a flat slab, 3 ft thick in places and supported on flared head capitals to the columns. Above this the bedroom floors spanned between the external walls and the corridor walls.

Curtis Green's plan (*Fig. 5.5*) differed very little from Williams'. The principal difference between the original design and the building that we now have is in the treatment of the elevations. Williams gave his design a definite horizontal emphasis, as surely Thomas Wallis would have done if his factory designs are any indication. This is particularly noticeable on the model where Williams has the windows wider than their height and the floor slabs projecting round the entire perimeter. However, in the perspective drawing the projection of the floor forms balconies on the long walls while on the wings the projections were limited to the length of the window opening. The horizontal lines are carried round the curved corners of the building with curved windows. Large amounts of glazing were used at the ground floor, particularly for the Park Lane elevation where the wall was stepped back with curved glass. At the rear, the half landings of the fire escapes projected at the ends of the two long wings and were glazed from the floor to the underside of the slab so that they appeared as glass tubes.

The construction of the walls, which Horder had failed to grasp, comprised a permanent shuttering of reconstituted stone behind which the concrete was to be cast. The inside face was cork lined, a device that was used for other concrete buildings of the period. In reporting the completed building, *The Architect and Building News* noted that '*[the facing] consists of precast concrete units having a marble aggregate [a terrazzo] which takes a polish*'[108] (*Fig. 5.6*).

Green's adaptation stripped away the modernist features of Williams' design and changed the horizontal emphasis to a vertical one. This was achieved by removing the horizontal projections, except for the balconies on the main elevations, and by changing the proportions of the windows, which would have had the effect of altering the light within the rooms. A restrained cornice was applied at the roof level, typical of Green's stripped classical style and a heavy string course two floors below that to divide off the 'attic' floors. The glazed walls at ground floor level were removed completely and were replaced with walls that had window openings corresponding with those above. The convex curves at the corners were changed to concave curves instead, so that the continuity of the walls was broken. This effect was then reinforced by infilling these convex forms with bronze and glass oriel windows from the second floor to the string course.

Had Williams been allowed to continue with the design, what would we have had? Robertson in his *Modern Architectural Design* used the Dorchester to illustrate how framed structures were liberating the plan;[109] Leathart in *Style in Architecture* referred to the Dorchester as an example of the use of permanent shuttering to form a concrete wall in his chapter on the Modern Style.[110] Despite the fact that the completed building was a stylistic compromise forced onto Green, it was generally well received by the architectural and engineering press. To the moderates, the building represented the true modern spirit in architectural design, while still retaining links with the past and displaying a national identity.[111] Modernists in Britain were disappointed. For them it was a wasted opportunity for Britain to proclaim its acceptance of the Modern Movement by building a large prestigious example of functionalist architecture. Richards, for example, described Green's adaptation as *'a genteel period piece which looks the compromise it is'*.[112]

Daily Express

Williams' role in the Daily Express building in Fleet Street, London, did not provoke as much publicity as the Dorchester largely

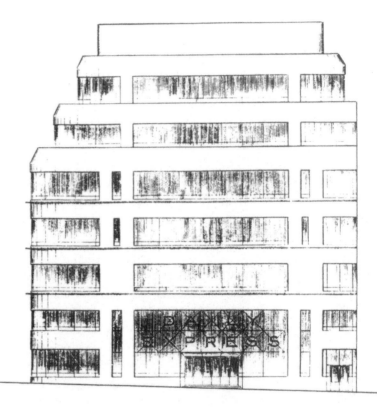

Fig. 5.7. Fleet Street elevation of the Daily Express building proposed by Owen Williams.

Chapter Five
Dorchester Hotel and Daily Express

because the original architects were retained by the client, even thought they, like Curtis Green, were working on a building conceived by Williams. Apart from which, a newspaper building in Fleet Street was probably not as sensitive an issue as a first class hotel in Park Lane. But even though Ellis and Clarke were officially credited with its design, the all-important structure belongs to Williams and it seems probable that he made a significant contribution to the decision to use glass cladding. The initial drawings that he produced between November 1929 and February 1930 included sketch elevations that have some similarity with those actually built (Fig. 5.7) even though it is clear that this idea was developed considerably by Ellis and Clarke. These elevations have large areas of glazed wall quite different from the kind of thing that Ellis and Clarke had produced, although they had built simple frame buildings with large windows much like most other industrial buildings of the time.

When Chermayeff reviewed the completed building he was fulsome in his praise of Williams' contribution to the design but failed to give a clear idea of just what this was. He wrote that:

'...the ingenuity of Sir Owen Williams' design in the basement was to be seen only in the early stages when one could obtain an uninterrupted view of the 58 ft spans between the stanchions required for the printing processes in the basement, where the concrete looked surprisingly light to be carrying the whole superstructure above.

It was the engineer's ability to give the basement a span 2 ft wider than the floors above, which was one of the deciding factors for the adoption of concrete construction in preference to steel.'[113]

Neither did the *Architectural Review* illustrate the building very well and it was left to the *Architect and Building News*[114] to produce the best known published section of the building, although even this only served to suggest that the principal structural issue was the 2 ft increase in span for the basement level (*Fig. 5.8a*). This section manages to obscure the critical change of span at the ground floor by being taken through the conveyor for the paper rolls. The caption reads 'An eccentric load on the left-hand post in the basement is transmitted by a reinforced concrete box girder to the retaining wall, which thus has an upward pull on it'. Perhaps

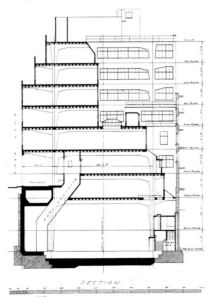

Fig. 5.8a. Published section of the Daily Express building (from *Architect and Building News*).

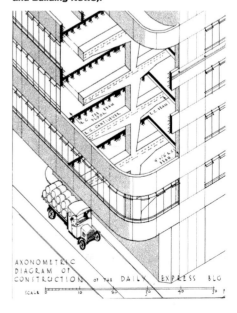

Fig. 5.8b. Explanatory diagram of the structure of the Daily Express building (from *Architect and Building News*).

Fig. 5.9. Reinforcing for Owen Williams' structure underpinning the original Daily Express building in Shoe Lane.

so, but this hardly explains the most significant feature of the structure and neither does the rather better diagram (*Fig. 5.8b*). In fact, the structure that must have secured Williams the job was the one he provided to enable the existing building to be modified, not the structure that he provided for the extension even though that too is an ingenious piece of engineering.

Elllis and Clarke produced survey drawings of the existing building during 1927 that show the arrangement of the existing steel frame structure that was to cause the problems with the press layout. The site had not been completely regular so this was not a simple orthogonal arrangement of columns and beams and the planning within the restricted site was hardly anything that one would have been proud of. What was now needed here was something to adapt this rather undistinguished building to the new extension and preferably something that involved a change in the layout of basement columns. Williams' solution was literally to cut through the existing steel structure above the basement and carry it on a new reinforced concrete structure. What he provided in the basement was essentially a giant reinforced concrete table on which the existing steel structure was to stand. This 'table' was a giant portal frame comprising a slab 3ft 6 in. thick at the

Chapter Five
Dorchester Hotel and Daily Express

Fig. 5.10a. View from under the structure that supported the original steel frame showing the tapering legs and the underside of the slab.

mezzanine level and supported on eight legs. The scale of this structure can be appreciated from the progress photographs that show the massive reinforcement in this concrete slab with the bars bending down towards the columns below (*Fig. 5.9*). The existing columns are visible in this photograph. When the concrete had been cast, the steel columns below it were then cut off. Below the slab were the legs which were 6 ft thick at the top but tapered in both directions (*Fig. 5.10a*).

This structure was wider and longer than the array of columns that it had to support. If this looks like a bridge structure then that is hardly surprising; there is some resemblance between this and the design for the bridge at Shepherd Leys Wood on the Bexley Heath bypass (completed in 1927). The impression is that Ellis and Clarke's original engineers were unable to suggest a structure that would solve the planning problem of the basement. However, the plans that *Architect and Building News* produced hardly explained the Owen Williams solution and one wonders whether anyone clearly recognised what he had done. The ground floor plan they produced (*Fig. 5.10b*) showed the column layout of the new frame clearly enough and the plans of the steel columns of the original structure. However, the basement plan that accompanied this repeated the same steel frame column layout and did not show the reinforced concrete supports that Williams had built.

In some ways this was rather like the structural arrangement at the Dorchester. There, also, a structure of closely spaced columns for the upper floors had to be carried across a much wider span below. In both cases the solution was to build a deep slab from which to carry the structure of the upper floors. The difference between the two was that, at the Dorchester, Owen Williams had control over the planning of the building above, whereas, at the Daily Express he had simply to accommodate what was already there. New steel bases were bolted to the existing 10 in. x 8 in. columns, which Williams estimated as carrying 70 tons, and these then transmitted the load to the concrete slab – the top of the 'table'.

Although drawings from Williams' office survive showing the reinforcement in this slab, the scale of the structure can be more clearly appreciated from the progress photographs. Of course, the result of this adaptation of an existing building is hardly like the nice, clean and logical looking structure of the extension. It has all the messiness that one often finds in such adaptations – hardly the thing to appeal to a Modern Movement architect. Perhaps this is why Chermayeff said nothing about it. However, he may also have had some difficulty in understanding the nature of Williams' intervention in the original part of the building because the floor levels of the two parts did not correspond.

We can of course assume that the planning and structure of the remainder of the building, the extension, followed from this solution for the existing structure. Bringing the wide space for the presses through into the extension would have led to the use of a series of portal frames there and the logic would then be to carry these up through the building.

This led to five transverse frames at 24 ft

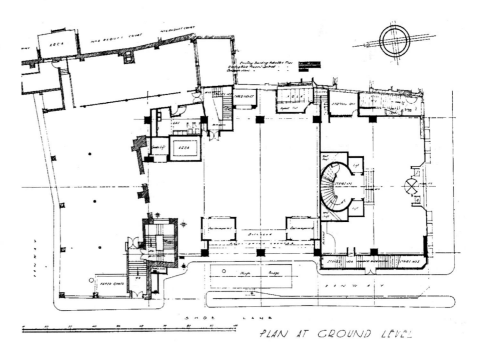

Fig. 5.10b. The published drawing of the Daily Express ground floor plan.

Chapter Five
Dorchester Hotel and Daily Express

centres and with a span of 56 ft 6 in. (*Fig. 5.8a and 5.10b*). Each of these comprised a pair of tapering columns at each floor supporting haunched beams and cantilevers that projected 18 ft beyond the column centre line on the Shoe Lane side and 13 ft on the other side. On the Shoe Lane side the cantilever was essential in order to provide an unloading bay for the paper deliveries but the result is a dramatic overhang (*Fig. 5.11*). These beams were 3 ft deep and at the time *The Builder* commented on the unprecedentedly low depth-to-span ratio of these beams, although the cantilevers would have provided some assistance here.[115] Between the frames were 12 in. deep beams at 2 ft centres cast integrally with the 3 in. floor slab to form a series of T beams (*Fig. 5.13*). The drawings also include calculations for a rather unusual structure above the fourth floor, where an inclined column was needed (*Fig. 5.14*). Williams was to use such sloping columns in his building for Lilley and Skinner.

Nothing of this structure can be seen from the outside because the frame at the Fleet Street front was handled quite differently and was made subservient to the requirements of the glazing. Paired, rectangular columns, set within the main structural grid, support deep rectangular spandrel beams forming the backing wall for the glazing. These are clearly expressed on the Owen Williams' proposed elevation but the paired columns are each concealed behind a single band of vitrolite on the Ellis and Clarke design that was built (*Fig. 5.12*). One suspects that this rectangular frame was a requirement of the architect in order to facilitate a simple rectilinear glazing pattern that expressed the position of these columns. The beams are then carried round to the Shoe Lane elevation where they serve little or no structural function – except perhaps to add dead weight to the ends of the cantilevers.

Fig. 5.11. Daily Express at night showing the cantilever over the Shoe Lane delivery dock.

Fig. 5.12. Principal elevation of the Daily Express extension.

Glazing

The external treatment of this building was unique in Britain at the time. The set back of the frames from the Shoe Lane elevation allowed the glazing above the ground floor to be continuous, alternating with bands of black vitrolite panels framed in Birmabright strips, a type of aluminium. The strips of opening windows were set back slightly from the black surface with narrow aluminium sills.

What is to be explained is how the architects produced something that was quite different from their original proposal and from all their other work. The inference is surely that the idea either came from Williams or from the Beaverbrook organisation itself, anxious to have a modern image for the building. This would certainly have enabled the paper to distance itself from the rather stuffy image presented by the neighbouring *Daily Telegraph* building only recently completed. The elevations among the engineering drawings (referred to above) are clearly design drawings from Owen Williams' office and not copies of the architects' suggested treatment of the frame. Also contained in the correspondence is a letter from Williams (27 April 1934) requesting out-of-pocket expenses and including the item 'Perspective'. If this elevational treatment

Fig. 5.13. Construction of a floor in the Daily Express extension showing a floor before concreting.

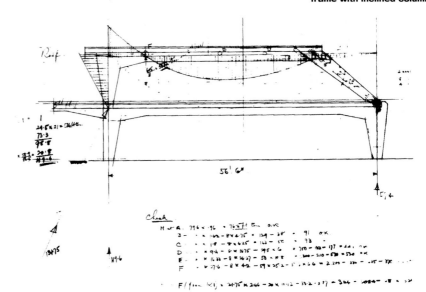

Fig. 5.14. Owen Williams' drawing and calculation for a frame with inclined column.

Chapter Five
Dorchester Hotel and Daily Express

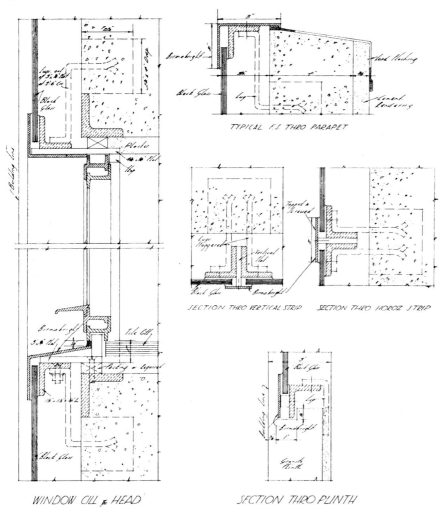

Fig. 5.15. Ellis and Clarkes's desing for the Daily Express glazing published in *Architectural Review*.

had been conceived by Ellis and Clarke, they, not Williams, would have commissioned an artist's perspective.[116]

The Owen Williams elevations certainly have some of the elements of the final design. He curved the corner, just as he had with the Dorchester, allowing horizontal bands of glazing to run continuously round the two elevations without interruption. However, the concrete frame of the Fleet Street elevation and the spandrel walls of the Shoe Lane façade were frankly expressed with infill glazing between and the drawing suggests projecting floor slabs with the glazing coming down to them. Williams also has the upper floors battered to line up with the geometry of the setback — the latter required to conform to LCC regulations. The impression is of a concrete and glass façade rather than one completely of glass. Such an arrangement would not have been unique at the time, having already been used by Thomas Wallis. Perhaps then it was Ellis and Clarke who suggested cladding the concrete with the black glass. However, what must surely be significant in the development of this wall was that Pilkingtons had acquired the Vitrolite Construction Company (Europe) Ltd in 1932 and reformed it as British Vitrolite.[117] They would have seen this building as an opportunity to promote their material and

Fig. 5.16. Design perspective of the Manchester Daily Express building.

are likely to have had a hand in the development of the detailing (*Fig. 5.15*).

Nothing like this was to be seen again until Belluschi's Equitable Life building in Portland, Oregon, although that used aluminum panels to express the structure whereas the Fleet Street building used Vitrolite. The two later buildings for the Daily Express, at Manchester and Glasgow, use a very similar elevational treatment in combining glass, Vitrolite panels and Birmabright but the construction details are different. Owen Williams was commissioned directly for these buildings with no involvement by Elllis and Clarke. By this time the paper presumably saw this

Chapter Five
Dorchester Hotel and Daily Express

dramatic glass wall as giving them a recognisable company image and would have wanted to have much the same. Moreover, at Manchester the site had a long frontage onto Great Ancoats Street but was on a corner site enabling paper deliveries to be round the back (*Fig. 5.16*). Glazing to the ground provided a view of the presses, an effect that was particularly dramatic at night (*Fig. 5.17*).

It is rumoured that the architects were rather miffed that Williams used their design details but in fact the glazing details at Manchester were much simpler than those used at Fleet Street. Here there was no set-back for the clear glazing with the black Vitrolite being in the same plane (*Fig. 5.18*). At Manchester, Williams used a flat-slab structure (see chapter 6) but there are columns within the walls expressed by vertical strips of Vitrolite as at London. But with the columns more widely spaced and the building having a longer frontage, it has a greater horizontal emphasis.

There is an even greater horizontal emphasis at Glasgow. The corner site at Manchester had allowed deliveries to be at the back but the paper's building at Glasgow had buildings either side so that deliveries and dispatch had to be at the front. The covered loading dock of the London building provided by cantilevering out over Shoe Lane, the side street, here had to be provided by cantilevering the building along its Albion Street frontage (*Fig. 5.19*). With the building cantilevered over the loading dock, there were no columns to express so that the building has simple horizontal bands of clear and black glass, the only vertical elements being at the ends. Although Williams used a flat-slab structure at Manchester, the structure at Glasgow was similar to that used at Fleet Street — 3 ft wide cross-frames at 21 ft centres span, 44 ft over a three-storey high press hall running along the rear of the building. The span is 39 ft over the

Fig. 5.17. Night view of the Daily Express building with the presses clearly visible.

publishing space with a 19 ft cantilever over the loading dock (*Fig. 5.20*).

The Fleet Street building appears to be a genuine collaboration between architect and engineer, with the engineering solution taking the lead. One may also presume that while the strong horizontal emphasis given to the building was initially Williams' idea, the final detailing was worked out by the architects, possibly with some collaboration by Pilkingtons or Crittalls, the window suppliers. Certainly it would have been the architects that would have removed the unconvincing battering of the upper floors and given the building the strong clean lines that it has. Surprisingly, acclamation came from across the architectural profession. Charles Reilly compared it favourably with its 'overdressed neighbours'.[118] It was even received as a well-considered piece of architecture by moderates, such as Howard Robertson and Goodhart Rendel, although it was thought too idiosyncratic to be a precedent for the street architecture of the future.[119] One wonders if their response may have been different had the building been accredited to Owen Williams rather than Ellis and Clarke. Naturally, the modernists could find no praise high enough, proclaiming it Britain's first large-scale example of modern architecture. Perhaps the most influential commentary was that written by Serge

Fig. 5.18. Manchester Daily Express building.

Chapter Five
Dorchester Hotel and Daily Express

Fig. 5.19. Glasgow Daily Express building — design drawing.

Chermayeff for the *Architectural Review* who was able to contrast it with the Daily Telegraph's new building in Fleet Street. Comparing the two, he wrote (first of the Daily Telegraph):

'Nothing in the stone-faced elevation gives one a clue as to the function of this structure except the letter of the name…The Express is quietly elegant in tight fitting dress of good cut which tells with frankness and without prudery of the well made figure wearing it. It commands admiration and respect from the onlooker, who must needs remain ignorant and indifferent to whatever charms and horrors are hid behind the upholstery of the Telegraph.'[120]

Crediting the architect — structure and architecture

In spite of his major contribution to both these buildings, Williams' role has never been fully credited. In a sense, architecture seems to have been regarded as something that was no more than skin deep so that Curtis Green has always been credited with the Dorchester and Ellis and Clarke with the Daily Express. Not even the engineering press credited Williams with the structural design of the Dorchester, regarding Considère and Partners as responsible for what was proclaimed as Britain's most advanced example of reinforced concrete design.[121] That Green himself never credited Williams with the design of the building is more distressing when his only role was to restyle Williams' elevations. Green already had a well-established reputation, which would hardly have been discredited by admitting his minor role in the project. Or perhaps he did not regard it as minor. The view of many British architects was that their prime function was as stylists. Williams took a critical view of this, claiming that there are two kinds of racketeers, *'those who get out their guns and those who get out their elevations'*.[121a] In that sense, the history of the Dorchester Hotel is a perfect illustration of Williams' position: that engineering principles were the determining criteria of architectural form and that architecture had degenerated to a concern for stylistic effect.

Williams himself tried to correct the situation in a letter to *The Times* in 1956. When an article dealing with a proposed extension to the hotel had referred to Curtis Green as the original architect, Williams wrote:

'I was appointed and did in fact plan and design the original hotel and it was built substantially in accordance with my designs, although during construction, owing to an insoluble disagreement on decoration with the clients, I resigned and thereafter the firm mentioned by you became the architects for the completion.'[122]

What should be noted from these episodes is the very different role played by the client in each. At the Dorchester, the clients had cold feet over the Williams designs and in the event preferred an architect who would give them something more conventional. At the Daily Express, the clients put their faith in Williams and somehow managed to have a firm of architects, who were not otherwise known for their modernist approach, collaborate with him in producing one of the outstanding modern buildings of the decade.

Chapter Six
Flat-slab buildings

Chapter Six
Flat-slab buildings

Williams' roles in the Dorchester and the Fleet Street building for the Daily Express have been largely forgotten, instead he is largely known for his building at Beeston for the Boots Company, the 'New Packed Goods Wet Building', or the so called 'Wets' building. This was his first architectural commission and, as noted earlier, not a particularly unusual one for an engineer to obtain. What is unusual is the use that he made of this opportunity. Even though it was followed by other successful industrial buildings by him during the 1930s, it was the project that is most commonly associated with his name and with his functionalist approach and was the one that received the greatest publicity. This building, like some of the others that he built, relies upon the use of flat slab-construction. Again, while this had fairly recently come into common use in Britain it was not a particularly unusual form for factory construction. What was special about this building was its overall planning and the way in which Williams combined the use of the flat slab with the glazing.

This building was one of his finest achievements, having a clear overall planning concept and well-integrated engineering details. At the time of its completion in 1932 it was highly acclaimed in the architectural press, represented by some critics as prophetic of the type of architecture that would become universally dominant, as designers returned to the sanity of 'science, reason and order'.[123] From the day of its completion its reputation has remained untarnished and it is now regarded as a building of seminal importance in the history of modern British architecture.[124] However, neither contemporary journal articles, nor histories of the period, have placed the factory in its true context, as a structure the planning and construction technique of which had clear precedents. These went back to pre-First World War factory design in America and to ideas that were already being imported into Britain. By overlooking these it appears as an unprecedented flat-slab building created by an engineer whose functionalist imagery was influenced by the continental Modern Movement. It was neither of these but that we can find precedents for its design is not to detract from Williams' achievement.

Factory planning

It was in 1927 that Williams first spoke about his revolutionary vision for the modern factory. In a lecture to the Art Workers' Guild he said:

'The object of the factory builder should…be fitness for purpose at minimum cost in a combination with complete flexibility for replanning and alteration:

"Fitness for purpose" I would like to regard very radically. For example, I would challenge the necessity for floors in the vast majority of factory buildings…Actually the factory building is the shell surrounding a process, and I venture to say that many processes are hampered by the imposition of floors. Once eliminate that conception of a factory and the process would take on a new efficiency. The factory can be likened to, for example, a colossal typewriter, but weatherproof and containing stages for its workers. The worker should control volume and not floor area, and requires a niche and not a surface…

I can picture the factory of the future as a great single span shell housing a vast machine with its workers dotted about in no way that can be related to definite horizontal planes or floors.'[125]

This view of the factory was not unique to Williams. No one who has seen Fritz Lang's *Metropolis* (1926), can fail to be reminded by this quotation of the factory scene in that film. The difference was that to Lang this image was one in which modern production processes had dehumanised the worker and reduced him to a mere part of a machine but to Williams it represented a beautiful model of efficiency that the Boots building came closest to realising.

At the time that he was designing the Boots factory the company was owned by The United Drugs Company of America[126] and we might expect this company, as its

Fig. 6.1. Plan and section of the Boots Wets building.

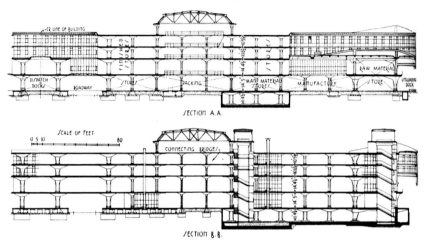

Cross sections through the lines A-A and B-B on plan

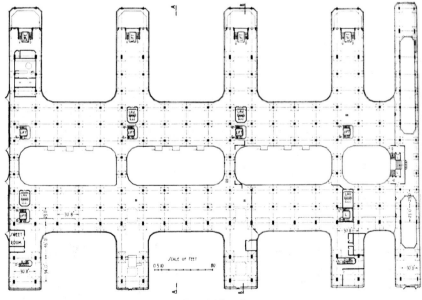

The third-floor plan

owners, to have had some influence on the organisation of production and hence the final form of the factory. The brief was developed in consultation with Boots' chief engineer and the Works Planning Committee, which had been established especially for this purpose. Through this consultation Williams was able to gather essential data on the production processes, machinery and workers' requirements. What was agreed was that the two types of pharmaceutical products — the 'Wets' and the 'Drys' — should both be produced in one large complex but that the construction should be phased with the Wets built and put into operation first and the Drys constructed as a second phase. Each of the two processes was also to include both manufacture and packing, with materials delivered at one side and the completed packaged products shipped out from the other. The result was that the building was conceived in two halves, handled so that raw materials were delivered to the outside of the building while the loading bays for the finished products were to be at the centre between the two production plants. Because transport at the time was by rail, this meant that rail tracks were to be brought through the centre of the building

In the event only the Wets portion was built as originally conceived (*Fig. 6.1*) and the Drys section was built as a separate and

Chapter Six
Flat-slab buildings

quite distinctive building (see chapter 7). It is not clear why this occurred. It may have been that by the time Williams was commissioned to do this, the company had reverted to British ownership and the new owners had different ideas. It could also be because lessons had been drawn from the operation of the Wets section that suggested a different method of organising the production process. Whatever the reason, what we now have is only half of the original plan with the present north elevation and its shipping dock initially intended as the centre line of the complete scheme. This accounts for the projecting floors and roof and the rather curious looking canopies on the south side (*Fig. 6.2*). This construction photograph, with the glazing being put in, shows most clearly that this was intended to be extended but, apart from the completed glazing, this elevation looks much the same today. Thus, in looking at the present building we have to make allowances for the unbuilt extension but it is possible to assess the overall design

Fig. 6.2. South side of the Wets building under construction.

Fig. 6.3. Interior of the atrium packing hall of the Wets building showing the chutes in use.

and to see how it was received at the time.

The layout of the building was conventional in one respect: it had the administrative offices as a screen across the front with the production and storage areas behind — but beyond this nothing was conventional. No other factory of the time had the railway running through and under the principal elevation. These tracks were to lead to the central loading docks that resulted from the basic layout adopted. Behind and parallel to this long loading bay were the packing halls and it is the packing hall of the Wets side that now forms the central feature of the building as completed. This is a vast, four-storey high, atrium space, separated into four sub-units by three narrow connecting bridges at each floor level joining the storage floors to either side. On either side of this space are a series of four-storey wings. The longer external wings on the 'outside' of the building had production on the ground floor with the upper floors used mainly for the storage of raw materials.

Raw materials arrived at the unloading dock to the south and were carried to the upper floors by the travelling cranes. From there the materials were fed down onto the production lines running across the width of the building with the finished Wets goods emerging on the southern side of the main packing hall. Packaging materials were fed to the packing lines from the upper storage floors by long gravity chutes (*Fig. 6.3*). Along the northern edge of the packing hall 28 elevators lifted the packaged goods to appropriate positions on the four-storey finished goods stores, the shorter wings of the building. Here they were sorted prior to dispatch by train and lorry on the northern side of the building. The railway line itself cut through the centre of this northern portion of the building and was ventilated by vast holes carved out of the flat-slab structure.

The use of a multi-storey building with a single storey ground floor was a common enough planning solution and was illustrated by Kahn in his book on factory design (*Fig. 6.4*).[127] It produced a compact building with the inclusion of single-storey central space allowing light into a deep

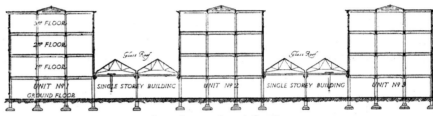

Fig. 6.4. Moritz Kahn's drawing of a factory based on the 'unit principle' with interior light wells.

Chapter Six
Flat-slab buildings

plan building. Here it made sense because a large proportion of the floor area was occupied by storage and grouping this round and above the assembly floor facilitated the handling involved. What was unusual was having this central space as an atrium, although Auguste Perret had previously used this device in his Esders clothing factory in Paris (1919) and it had been illustrated earlier in the *Architects Journal*.[128] However, it is also possible that this idea came from Ford's pre-First World War buildings. For example, a large atrium space was used in conjunction with flat-slab construction in Ford's Detroit plant around 1910. It was Ford who introduced the atrium and gravity feed method of car production before the First World War, immediately prior to the development of his assembly line concept. These earlier American atrium schemes contained long tall spaces, very similar to that at Boots, along which the assembly line was arranged. Components were fed to the assembly floor either by gravity chutes or travelling cranes.

Williams himself would probably have known of this, illustrated as they were in *Engineering News Record*. Photographs of the interiors of these buildings, with their atria surrounded by cantilevered balconies at 4–5 floor levels, have some affinity with Williams' packing hall at Boots.[129] The main difference between them is that Williams arranged the production lines transversely and not longitudinally in these spaces. This atrium arrangement might have been used at Boots at the request of the American client. It is difficult to justify such a volume on a purely functional basis merely to provide roof lighting, impressive thought the space is, because the heating costs are phenomenal.[130] Neither are the gravity chutes within the space sufficient justification because they could have been arranged as vertical spirals through the floor slabs, as others had done elsewhere. On a human level also, there is little doubt that while the packing hall is awe-inspiring it is also dehumanising. It is a dominant space whose scale and efficiency reduce the individual worker to insignificance. The spectre of Fritz Lang's image brought to life.

Most architects grouped lift shafts, stairs and toilet accommodation in tightly planned blocks along the elevations of their factories. Instead, Williams positioned them in strategic positions in the centre of his building (*Fig. 6.5*) where they were structurally independent of the main flat-slab arrangement and expressed only as projections above the roof. Thus, the façade design and the ancillary accommodation were all made subservient to the simplest possible arrangement of the flat-slab structure.

Fig. 6.5. Stair structure of the Wets building.

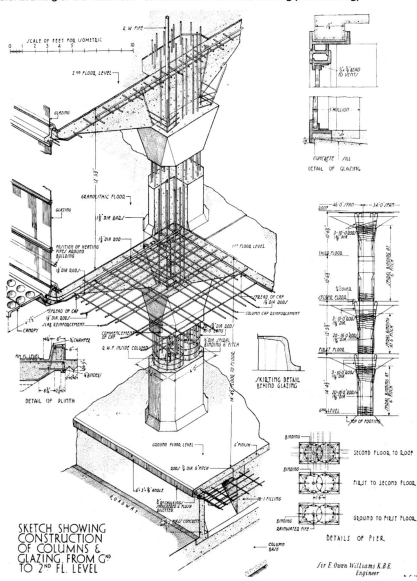

Fig. 6.6. Drawing of the 'mushroom' construction at the Wets building (from *Building*).

Structure

The flat-slab construction, on which this building was based, was a technique that had been developed both on the continent of Europe and in American but Williams' sources were probably the latter and a brief summary of its introduction to Britain is useful here to place it in context.[131] It comprises nothing more than a flat floor plate resting on columns without any downstand beams, its structural behaviour depending upon the arrangement of reinforcing within the slab (*Fig. 6.6*). 'Mushroom' heads are required at the tops of the columns to deal with the shear forces. Its advantages for factories are that it simplifies the shuttering of the concrete and makes the installation of service runs and machine shafting very much simpler. Also, the absence of beams, especially edge beams, can improve daylight conditions within the building. The technique was developed in the United States where it came into widespread use for industrial buildings because its structural advantages are greater with the higher floor loads associated with factories and warehouses. In spite of these advantages, its introduction into Britain was hampered by the LCC's reinforced concrete regulations, which were also used by other authorities in their by-laws.[132] Therefore, it tended to be used in structures that were exempt from by-law control. However, both architects and clients continued to show an

Chapter Six
Flat-slab buildings

interest in its use, particularly, American clients building in Britain and it was gradually accepted by the mid-1920s.

It was used for factories in Britain built for the American Shredded Wheat Company's factory in Welwyn Garden City (Louis de Soissons, 1926–27)[133] and at the same time by Wallis Gilbert and Partners for the Wrigley Factory, Wembley. The structures for both of these were designed and built by the Trussed Concrete Steel Company. During the construction of the latter *Concrete and Constructional Engineering* had pointed out that: '*[The owners] knew that the cheapest and quickest method for such a building was the mushroom type of construction*'.[134] The favourable reviews that such buildings received have already been commented on and the use of flat-slab construction was being actively encouraged at the time by the *Architect and Building News*. By the time Owen Williams came to use it at Boots, others in Britain had already used it to some effect.

At the same time that Williams was building the Boots factory, the Indented Bar Company was putting up Viyella House as a flat-slab structure for the architect F. A. Broadhead and in the same city.[135] Like the Boots factory, this building used extensive glazing between narrow strips of cantilevered floors. Its elevations, however, were less convincing. Although the building did have deep windows, a low upstand wall was embellished with mouldings clearly drawing attention to the floor divisions. The entrance was also marked with a group of 'pilasters' rising to the third floor framing a two-storey high entrance feature (*Fig. 6.7*). If the engineer's framework had been left undecorated in this way it would have more closely resembled the Boots factory.

Williams conceived the whole building, including the office block, as a four-storey, flat-slab structure arranged on a rigid grid layout, with light and ventilation wells carved out of it as appropriate for the ground-floor production and packing processes. The flat slab was not essential to the planning of the production process although the absence of columns at the edge of the galleries did aid movement and flexibility in the positioning of

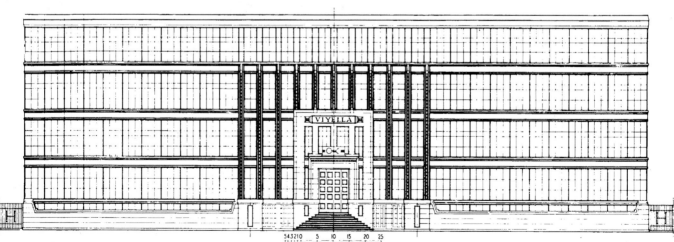

Fig. 6.7. Principal elevation of the Viyella factory, Nottingham by F. A. Broadhead.

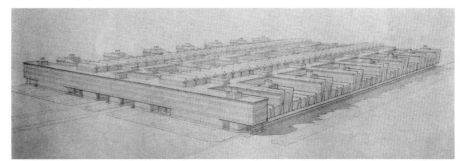

Fig. 6.9a. Design proposal for the Boots factory — not as built.

Fig. 6.8. Cantilever over the unloading dock — one of the crane rails is clearly visible.

Fig. 6.9b. Design perspective of the Boots factory.

the chutes. It was an appropriate solution for heavily loaded floors, although it was not needed here for under-floor service runs nor for lighting. Naturally, it would have facilitated the daylighting of the office areas although it was not otherwise required for that part of the building, especially not at the same size because of the lower floor loadings. One has the impression that it was used there as well in order to give an architectural consistency to the whole scheme. Indeed, the form of structure that Williams used was rather heavy compared with other examples of flat-slab construction. It was dramatically expressed because of the long (30 ft 8in. x 23 ft) span of the grid and the size of the column capitals, which were much larger than usual. His mushroom heads are not the simple inverted cones normally added to the tops of circular columns. Instead octagonal columns are first brought to a square to carry shallow, wide-spread inverted pyramids (*Fig. 6.5*).

An essential feature of the design was that the structure should cantilever out over the unloading dock (see *Fig. 6.8*) and this was the only substantial part of the building where Williams departed from flat slab construction. Large double cantilever beams projecting above the roof were used to support both heavy roof construction and a travelling crane, itself supported from hangers suspended from the ends of the 40 ft cantilevers. In the earlier perspective the crane hangars were strongly expressed as projecting fins on every gridline (*Fig. 6.9a*) This made the ends of the four-storey

Chapter Six
Flat-slab buildings

Fig. 6.10. Perspective drawing of the principal façade of the Boots factory.

Fig. 6.11. Principal façade of the Wets building — compare with Fig. 6.10.

wings quite different from the long four-storey office block. But Williams seems to have been striving for as much uniformity as possible in his treatment of the walls and in his later perspective, and in the final design, these hangers are reduced to a minimum so that the bevelled corners read as much the same on both the wings and on the office block (*Fig. 6.9b*).

Glazing

What was striking about this building, compared with others, was the combination of the flat slab with the almost continuous glazing. Just as there was no change in the structure between the production area and the administrative block on the front so, in the final building, the glazing of the office block is also the same as the production areas. However, examination of the early perspective drawings shows that the design was refined so that this similarity of treatment became clearer. A perspective drawing of the main façade suggests that Williams might have intended the building to have either a curtain wall, with the glazing continuous past the floor slabs (*Fig. 6.10*), or perhaps to have used much the same glazing arrangement as in the Daily Express building that was in the office at the same time. The complete glass façade was one of those ideas that was current at the time. Mies, in Germany, had proposed a number of glass-walled buildings, beginning in 1920, and Korn's book on glass in building had recently been published,[136] but it is difficult to know how aware Williams might have been of such ideas. However, he was surely aware of the recently completed Van Nelle factory (Brinkman and van der Vlugt, 1929) with its continuous glazing. In the event, Williams chose to have glazing between projecting floor slabs but with no spandrels so that there is an uninterrupted view of the interior.

Most architects at the time treated factory administrative blocks as a distinctive architectural unit. Even Brinkman and van der Vlugt, who also used a frankly expressed flat-slab and a glass curtain wall in their Van Nelle factory in Rotterdam, had separated the administration block and given it a distinctive architectural form. At Boots, however, the perspectives show similar ground floor treatment of the glazing, compared with the

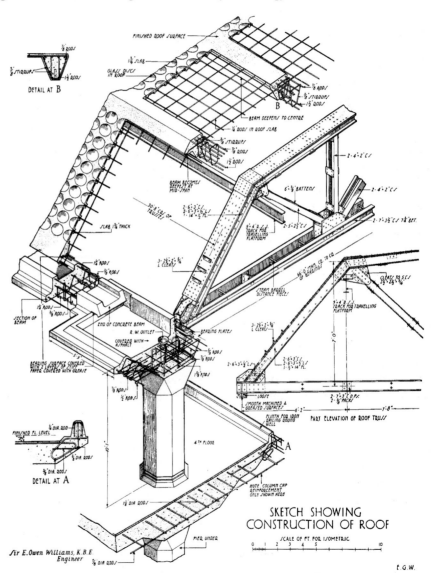

Fig. 6.12. Construction drawing of the atrium roof (from *Building*).

floors above, and the entrances marked simply by being deeply recessed within this otherwise plain façade. In the event, even these deep recesses were to disappear, the entrance doors simply being set back 3 ft at the ground floor (*Fig. 6.11*).

The other change that we see in the development of the design is in the arrangement for lighting the packing hall. The first perspective shows a combination of clerestory lights and simple conventional transverse roof lights. In the later perspective, the roof lights have gone and only the clerestory arrangement is clear. Eventually the atrium was roofed with steel trusses between which are glass blocks set in concrete (*Fig. 6.12*). It also seems that Williams intended having a screen of the same construction hung across the full length of the rail loading bay. However, the technology of the time was not sufficiently well developed to enable this to be done and there was apparently some consternation about the idea within the glazing industry. George Crabb, who was working for Owen Williams at the time recalls how he was 'buttonholed' by the proprietor of one of the concrete and glazing manufacturers, who opined that if Williams was to continue with this scheme he would ruin the industry. Crabb says that he passed the message on and a simple hanging glazed screen was used instead.[137]

Chapter Six
Flat-slab buildings

The result

With advice from the Works Planning Committee, Williams produced a masterpiece, organising the production process within a highly intricate three-dimensional arrangement. The confluence of this three-dimensional planning system was the main packing hall, the centre piece of the production process and a vast atrium space, cathedral-like in its proportions. No one can fail to be impressed by the visual dynamism of this space with its vast steel truss roof structure supporting a glass disk and concrete deck that provided effective diffused light at ground floor. At the same time, the unified treatment of the glazing, the complete absence of any spandrel walls and the view thus presented of the regular rhythm and bold mass of the structure behind cannot fail to impress the viewer. The importance of this structural device is that much of Owen Williams' reputation depends upon the positive response of the press to the Boots factory.

While we can trace precedents within the design that clearly influenced Williams, he greatly improved upon them by starting with the organisation of the factory and developing the design from the inside out. This was the first time in his career that Williams was entirely free to apply his own ideas unencumbered by the involvement of an architect. The result was an impressive structure whose forms and details were visibly determined by the function of the building and the most efficient use of the structural materials employed. But in the development of the design we can see that Williams was not oblivious to architectural needs, working to give the whole building a unity of treatment that is its hallmark.

Recognising the precedent for the glazed elevations in the Van Nelle factory, Charles Reilly commented:

'The factory for Messrs Boots in glass and concrete is of course not so thrilling as the Van Nelle tobacco factory at Rotterdam by Brinkman and Van der Vlugt in the same materials, but one can hardly expect an English engineer in his first experiment to equal two of Holland's best architects. Still it is a great step forward in this land of muddled, illogical factories.'[138]

Despite this faint praise, the Boots factory was otherwise regarded as the most impressive example of British industrial architecture of the time. Forty years later it was described as *'Britain's Crystal Palace of the 20th century, and the most advanced piece of industrial architecture in Britain before the Second World War'*.[139] At the time the architectural press received it as an unprecedented example of functionalist architecture in Britain, which proved that scientifically conceived structures could achieve great beauty. The editors of *Building* opined that it would have been impossible for an architect to produce a structure of this type.

'It is difficult for a trained architect to be of the true functionalist faith — his aesthetic training and temperament make it almost impossible. And thus it is hardly surprising that Britain's most outstanding functionalist building has not been designed by an architect at all, but by an engineer.'[140]

To Williams the building proved that his functionalist approach was effective and it provided him with the confidence to use the same philosophy in the design of other projects. To those committed to the modernist cause, the building represented the first large-scale example of modern architecture in Britain, providing evidence that only through a rejection of stylism and the adoption of functionalism could appropriated twentieth-century architectural forms be created. This was the 'reason and order' that *Architect and Building News* had spoken of. The moderates and the traditionalists within the profession were less enthusiastic, considering the building to be an example of engineering rather than architecture. In this way they could applaud it as a well-considered piece of engineering work, designed by an exceptionally gifted engineer, which, unlike the Dorchester, presented no particular threat to the status of the architect.

Fig. 6.15. Sainsbury's building, Blackfriars, as constructed — compare with Fig. 6.18.

Sainsbury factory and warehouse

Following the Boots factory, Williams made a couple of other essays in the use of the flat slab, one built and the other not. The first was a factory and warehouse for Sainsburys in Blackfriars (1931–33) for which this type of structure seems ideally suited. The site was rather irregular so that the long walls were not parallel. If a simple frame had been used the beams would have had to be of different lengths, whereas the flat-slab arrangement allowed a regular grid of columns to support a varying span of floor (*Fig. 6.13*). In the wide part of the plan the two rows of columns were set back from the walls while the narrow section had the columns incorporated into the wall. Here the span was so short that only columns in the wall were required and the mushroom heads were reduced to tapering cantilevers at 45° to the wall (*Fig. 6.14*). The span of the main area was such that an external wall frame was required so that the flat slab construction of the interior is not apparent. The elevations are divided by pairs of columns with solid spandrels under the windows (*Fig. 6.15*). These columns eventually support the edge beam of the roof, which is structured quite differently from the floors below. There slender columns support a wide rooflight using the same glass block technology as the Boots

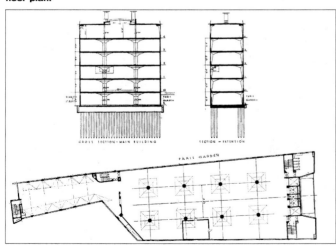

Fig. 6.13. Sainsbury's building, Blackfriars — floor plan.

Fig. 6.14. Sainsbury's building, Blackfriars — interior of the narrow part of the plan.

Chapter Six
Flat-slab buildings

atrium (Fig. 6.16). The structure here used beams above the roof (Fig. 6.17), the device that he had used for cantilevers at Boots. Externally, the striking visual features of this building are the staircase and window-washing cradle. The former is set in the long straight wall but marks the junction of the two sections of the building and is clearly expressed as a series of large steps. The rail to support the window-washing cradle is carried on a series of cantilevers that are at the same interval as the window mullions and one sees immediately that it is their size that provides the geometric order to the elevations.

This building has subsequently been adapted as an office block. Suspended ceilings have covered the heads of the columns and the cladding has been altered (Fig. 6.18). Another feature of the building that has now been lost are the supports for the window-washing cradle.

Williams was also to consider the use of a flat slab in his design for the Cumberland garage and car park (1932–34) but in the event a conventional post and beam structure was used. This is a pity because it was one of a number of very early car parking structures built in London in the inter-war period, some of which were featured in the architectural journals of the time. In reporting the completion of a parking structure at Piccadilly Circus in

Fig. 6.18. Sainsbury's building, Blackfriars, after refurbishment work — note the removal of the window-washing cradle supports.

Fig. 6.16. Top floor interior of Sainsbury's building, Blackfriars.

Fig. 6.17. Beams above the roof of Sainsbury's building framing the rooflights.

1929, an article in *Architect and Building News* had suggested the use of flat-slab construction for this type of building because of the deep beams that it had noted were otherwise needed.[141] Williams approached the problem from first principles. By considering the way in which cars moved and the different possible method of transporting them through the building, Williams arrived at three possible solutions. One involved the use of electric lifts, another a flat-slab structure with a spiral ramp and the third a split-level scheme with a post and beam structure; the one eventually built. In this arrangement, the front and rear halves of the building were separated by half a floor to provide acceptable gradients on the connecting ramp — a system commonly used today. However, the requirement of the local planning authority for a brick façade prevented any expression of the structural concrete.

The Pioneer Health Centre
There is little doubt that commonly held attitudes to the relationship between architecture and engineering made it difficult for Williams to acquire work normally within the preserve of the architectural profession. It was not that individuals within the profession directly prevented his acquiring such work, rather that the mindset of clients worked against him. To overcome these prejudices it was essential that he acquire 'architectural' projects to prove his design abilities were equally effective when applied to all types of buildings. Had he succeeded with the Dorchester this barrier might have been substantially removed at the outset. In the event, it was not until the mid-1930s that he gained commissions that enabled him to demonstrate his abilities. Of these, the Pioneer Heath Centre needs to be considered in the context of his other flat-slab buildings and was a demonstration of how effective Williams was with non-industrial projects. Its structure was also significant as the first non-industrial building in Britain to use flat-slab construction but, although reference was made to this in some architectural journals, this was not what caught the attention at the time. The building received particular attention because Williams' radical functionalist design principles seemed to respond, and even enhance, his client's radical social medical beliefs. However, once again his appointment was not without controversy.

The official name of the client was the Pioneer Health Centre Limited, comprising three directors: two doctors, Innes H. Pease and G. Scott Williamson, and the sociologist J. G. S. Donaldson. They had two interrelated objectives, to make a biological and sociological study of the working-class

Chapter Six
Flat-slab buildings

family unit and to provide leisure and health-care facilities for a specific working-class community in the belief that the true function of medicine was to preserve health.[142] The intention was to combine these in a recreational and health care setting to which families would subscribe (at one shilling per family per week) and where they would join in sports activities and receive regular medical checks and health education. For their part, the researchers intended to operate the centre rather like a laboratory in which they could undertake research, collecting essential sociological and medical data on the health and social development of working-class family units.

In order to assess its feasibility they had established a pilot project in 1926, using a private house in Peckham for a period of about two years. The result of this exercise were sufficiently encouraging for them to proceed, so that in 1930 they decided to publish their initial findings in a book to attract sponsors for a purpose-built scheme.[143] In the book they included initial design proposals for the centre, elaborated for them by E. B. Musman.[144] This was a simple symmetrical arrangement of central block with a grandiose entrance, quite out of keeping with the ideals of the clients, with two-storey wings on either side, all in the modern style. In 1933, when their financial position was sufficiently strong for them to seriously contemplate the design of a new building, the clients abandoned Musman's scheme and recruited J. M. Richards to formulate a more comprehensive brief by sketching out space allocations and organisational relationships. Richards, recently returned from Ireland, was contemplating entering journalism and recalls in his autobiography Williams' surprise when he visited the doctors' house in Peckham and recognised his once junior assistant already at work on the drawing board.[145] Richards seems to have been unaware that his preliminary sketches had been circulated by the doctors to a number of architects in order to obtain from them sketch designs and cost quotations. He was thus unaware of the stir that this caused within the profession.

One architect who received these sketches was Goodhart-Rendel. With them was a covering letter in which Innes Pearse requested confirmation that the proposed building could be erected for less than £25 000 and for each architect to supply a tender figure for its design and construction.[146] Goodhart-Rendel immediately wrote to McAlistair, secretary of the RIBA, to fulfil his professional duty in exposing the possibility of unprofessional conduct. Enclosing a copy of Pearse's letter and referring to the promoters of the exercise as cranks, he wrote:

'I need hardly say that I do not send this correspondence to you with the least intention or

Fig. 6.21. Entrance to the Pioneer Health Centre.

even desire of preserving the job for myself, but because I do feel that anybody who sent a reply, in a different sense to mine, ought to have his knuckles rapped.'[147]

The RIBA's competition committee was informed when it became clear to Goodhart-Rendel, via a letter from Pearse, that other architects had contravened the RIBA's code of conduct by submitting prices. However, neither Goodhart-Rendel nor McAlistair were able to discipline the successful tenderer. When they realised that Owen Williams had been awarded the contract, Goodhart-Rendel noted:

'I have heard unofficially that it is extremely likely that Owen Williams has undermined all

Fig. 6.19. Pioneer Health Centre — principal elevation.

Fig. 6.20. Pioneer Health Centre — play area under the main front of the building.

opposition to him and landed the job, over him we have no control.'[147a]

To many modernists of the time its combination of radical concepts made the Pioneer Health Centre something of a cause célèbre for the British Modern Movement, as it was regarded as a scientifically conceived structure with a socially progressive function — two essential requirements for truly modern architecture. The *Architectural Review* was particularly influential in projecting the building in this way.[148] However, the writer of its article was J. M. Richards who could not be described as wholly impartial, both because he had helped the client formulate the brief and was predisposed towards Williams' work through his earlier contact with him. Although the modernists understandably used the building to support their claim that the true function of the modern architect was to instigate social reform, the evidence suggests that the prime reason for Williams' appointment was that he offered to design and construct the building for a much lower cost than a number of architects who were invited to tender in what the RIBA regarded as a wholly unprofessional design and build competition. The letter sent out to the various architects with the plans shows that

Chapter Six
Flat-slab buildings

the client's main concern in their choice of architect was cost, with the type of structure being of secondary importance. Its clear implication is that had Goodhart-Rendel or any other architect submitted a lower priced proposal he would have been appointed even though his approach to architectural design might have been quite different from that of Williams. There is no evidence that Williams shared any of the social ideals of his clients. His contribution was to provide a built enclosure that responded as efficiently as possible to a predetermined brief and within the clients' budgetary limits.

Williams located his building at the north-eastern edge of a two-acre site in Peckham off St Mary's Road (*Fig. 6.19*). This

Fig. 6.23. Pioneer Health Centre — interior of the pool from the diving board.

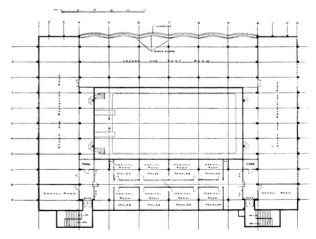

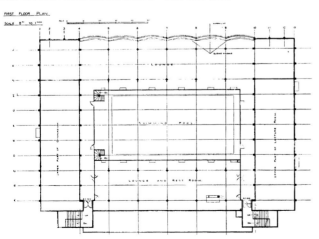

Fig. 6.22a. Pioneer Health Centre — first-floor plan.

Fig. 6.22b. Pioneer Health Centre — second-floor plan.

was done in order to preserve as much as possible of the site's south-westerly aspect. This area fronting the road provided open-air leisure facilities that included a running track, tennis courts and children's play area. The play area continued under part of the first floor (*Fig. 6.20*), providing an open covered play space adjoining a nursery in the building's south-eastern wing. To maximise the visual and physical contact between the internal and external spaces on this south-westerly face he placed the main entrance in the centre of the rear elevation, (*Fig. 6.21*) quite unlike Musman's original sketch design. Although the rationale for this decision is clear, it resulted in an unimpressive approach to the building, down a dark, narrow alley feeding a long rectangular foyer that directed visitors to the two staircases and corridor doors at either end.

In planning and structural terms, the building was organised as a symmetrical composition with four distinctive rectangular blocks grouped round a central, internal swimming pool (*Fig. 6.22a–d*).[149] The pool is the central feature of the interior occupying the entire height of the building and providing a focal point for the recreational spaces on the first and second floors the internal glazed walls of which look into or onto it (*Fig. 6.23*). The blocks round the pool were three storeys high, comprising two slightly projecting wings to the south-east and north-west, with insert blocks to the north-east and south-west. At ground floor level the two wings contained a gymnasium, nursery and lecture room, with the insert blocks reserved for machinery, changing spaces and covered playground. The first floor level was intended to be the part of the building where most of the users would socialise in a lounge area that filled the centre of the south-west frontage, overlooking both the internal pool and the outdoor sports facilities (*Fig. 6.24*). At the rear of the building, on the first floor, was the cafeteria. The upper parts of the gymnasium and lecture room occupied a significant volume of the first floor. The second floor was the only part of the building where a distinction was made

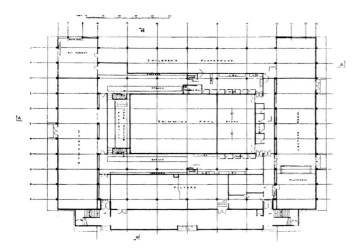

Fig. 6.22c. Pioneer Health Centre — ground-floor plan.

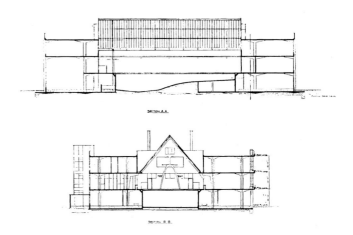

Fig. 6.22d. Pioneer Health Centre — cross-sections.

Chapter Six
Flat-slab buildings

Fig. 6.24. Pioneer Health Centre — main lounge area.

Fig. 6.25. Rejected design perspective of the Pioneer Health Centre.

between the staff's private quarters and the public spaces. The former were placed above the cafeteria to the north-east with the remaining U shaped floor area round the glazed roof of the pool occupied by library, study and recreation spaces.

The present form was not the only one that Williams contemplated, as can be seen from two perspectives that were prepared. The rejected design, clearly less satisfactory, shows the building far more rectangular in its appearance (*Fig. 6.25*). In the central block the eaves' line and edges of the floor slabs are straight and within their length have only four bays of balconies formed by zigzagging the glazing. The wings were then recessed behind these. This produced a rather uncomfortable step in the roof-line between the different elements of the building. Also, the side blocks were rather simply treated, with the glazing on the front elevation uninterrupted by columns. Of course, this plainly expressed the nature of the frame behind. In the side elevation the columns were broad and shallow rather than narrow and deep, and were slightly recessed behind the plane of the spandrel beams. The effect would have been to continue the horizontal emphasis right across the front and round the sides.

The changes, which would have involved some modification of the plan, banished the set back at the ends, so increasing the number of balcony bays to six (*Fig. 6.26*). The centre block was now differentiated from the wings by a change in the treatment

Fig. 6.26. Perspective of the final design of the Pioneer Health Centre.

Fig. 6.27. Night view of the Pioneer Health Centre.

of the edge of the roof slab and by introducing verticals into what could, structurally, have been the same plain elevation. At the same time, this change eliminated the glazing on the ground floor of the centre block allowing the play space to sweep under the building. It is clear from these two perspectives that Williams was not designing on the basis of the plan alone but was taking into account the overall massing of the building and the architectural relationship between the centre block and the wings. The arrangement, with the double height spaces for gymnasium and lecture theatre in the wings, stands out most clearly at night (*Fig. 6.27*). In the daytime the projecting columns emphasise the verticality of the wings in contrast to the horizontal emphasis of the central block with its projecting floor and roof slabs.

The external glazed walls were made up of folding glass screens, arranged on a zigzag plan form, which could be opened up in good weather converting the lounge into open balconies overlooking the external sports area. However, that angular plan form of the glazed screens has not been carried through to the elevations because the edges of the floor slab form a series of curves producing the overall impression of a series of domestic bay windows (*Fig. 6.28*). The development of these is clear from the two perspectives and may have been intended to

101

Chapter Six
Flat-slab buildings

Fig. 6.28. Pioneer Health Centre — bay windows of the front elevation.

give the building something of a domestic quality in keeping with its function. In both drawings, the zigzag of the glazed screen is set back at the centre of each bay so forming a permanent balcony area.[150] The subsequent change naturally throws this space into the room itself, perhaps a more sensible idea in the English climate. The one disadvantage of all this glazing round the perimeter of a relatively small building was that it produced problems of thermal control, which were more acute than those experienced at Boots.

Apart from the staff accommodation, arranged off a simple double loaded corridor, the subdivision of the interior was largely by glazed partitions, allowing visual contact between activities. This enabled the researchers to observe the activities of the families. The use of glazed walls round the pool, allowing views downwards and upwards as well as across, even overcomes something of the separation of the floors. Something of this can be seen from the view of the gymnasium from the upper floor (*Fig. 6.29*). The extensive use of glazing, which was carried through to the external walls, produced high internal daylight levels and a quality of openness that was one of the building's most notable attributes. The clients were particularly pleased with this latter quality. In their book they wrote:

'*The general visibility had continuity of flow*

Fig. 6.29. Pioneer Health Centre — view from the lounge into the gymnasium.

Fig. 6.30. Pioneer Health Centre — detail of the column supports for the flat slab.

throughout the building which is a necessity of the scientist. In the biological laboratories of biology and zoology the microscope has been the main and requisite equipment. The human biologist also requires special sight for his field of observation — the family. His new "lens" is the transparency of all boundaries within his field of experiment. Sixteen steps down from the consulting room he is engulfed in action which is going forward, and which, by the very design of the building, is visible and tangible to his observational faculties at all times.'[151]

The quality of openness was a direct result of Williams' concrete frame structure. A traditional masonry structure would have required extensive loadbearing walls that would have resulted in a more cellular closed plan form.

It was his use of a modified form of flat-slab construction that contributed most to the openness of the planning and the overall visual quality of the building. The use of flat-slab construction at Peckham could not be justified by the heavy floor loads that

normally made it an efficient form in commercial buildings. However, the set-back columns and floors cantileverd 8 ft beyond the column line ensured the uninterrupted glazing that maximised visual contact between the various activities, both within the building and with the outside. The system also provided the client with flat soffits, free from downstand beams that maximised the daylight and would enable alternative arrangements of the glazed partitions.

Because the loading requirement on the

Chapter Six
Flat-slab buildings

Fig. 6.31. Shuttering for a column head of the Manchester Daily Express building — note the cast in sleeve.

Fig. 6.32. Interior of Manchester Daily Express building during fitting out, showing sprinkler pipes through the column head.

structure was minimal, compared with an industrial building, Williams was able to reduce the column capitals to four tapered arms merging elegantly into the concrete floor soffit (seen in Fig. 6.30). This was an arrangement that he had used previously on Dalnamein Bridge. Normal mushroom heads would have been visually intrusive in this situation, where the room heights were much more domestic in scale. Thus, his modification of the flat-slab technique was more appropriate for the small-scale building and the relationship between the proportions of the spaces and the structural supports is entirely harmonious.

The similarities between this and the Boots building are too obvious to pass without comment. Both have flat-slab structures grouped round a light well; both have extensively glazed façades, largely as a consequence of the flat slab and the logical solution to the requirement for good daylighting; both have a minimal subdivision of internal spaces, although in both cases this is a direct result of the brief. Williams also applied the same rationale to the detailing of the two structures by having exposed concrete surfaces, except where specific conditions required otherwise, such as the tiling round the pool at the Peckham Health Centre. The use of plain concrete was, of course, normal for an industrial building, at Peckham it would have involved a saving of plastering costs, which must have been a factor in his ability to build at the price the client was willing to pay. Of course, many of these finishes appear crude and in complete contrast to the refined detailing associated with most public buildings of the time.

Was it that this was a direct transfer of his ideas from an industrial building to the recreational building which was simply fortuitous in producing a highly acclaimed successful result? There is at least one aspect of the building where the planning of the spaces appears to have been subordinated to the symmetry required by an efficient structure. This is at the second floor level where there are two identical spaces in the wing blocks, each complete with open fireplaces, and both labelled 'Study and Recreation Rooms'. Was there any justification for this duplication or is it, as one might suspect, that he felt compelled to provide a superfluous space in order to maintain symmetry? Rather than having been fortunate in having his limited architectural vocabulary succeed in this situation it could be argued that Williams was far-sighted in applying an industrial building form to this brief, even though it required a domestic-scale building, and that he demonstrated considerable skill in carrying this off successfully. This is surely how the Modernists would have viewed it. He was either inflexible and fortunate or a clear-thinking creative architect. Whatever view one takes, the success of the building is undeniable.

Postscript

This chapter requires what is essentially a postscript because Williams was to use the cruciform columns of the Pioneer Health Centre in the later Manchester Daily Express building and for the Boots Dry's building. In these buildings they were used with much wider spans and naturally much heavier loading. In the Drys building, the columns are on a 36 ft grid with 24 ft side spans, although it seems likely that some assistance is given to the latter by the external hangars (see chapter 7). At Manchester, the columns divided the building into three — a central span of 39 ft with side spans of 25 ft. The advantage of this form over the simple mushroom head normally used for flat-slab construction seems to have been that the services could be run much closer to the columns. Within the Wets building one can see how Williams anticipated vertical service runs by casting ducts within the mushroom heads but the restriction such large mushrooms would have had on the running of horizontal services has already been noted. Progress photographs of the Manchester building show how horizontal pipe-runs for the sprinklers system were accommodated close to the columns by casting tubes into the lightly stressed centres of the brackets (*Figs 6.31 and 6.32*). As well as allowing for the pipe runs through the column brackets, Williams also designed an office floor of removable ply panels resting on battens so that lighting and telephone cables could be run underneath. This service floor could be taken up and changed whenever there were changes in the office layout.

Chapter Seven
Long-span buildings

Chapter Seven
Long-span buildings

There is an expectation that an engineer comes into his own when dealing with long-span buildings. If so, then how does Owen Williams stand up as a designer of long-span structures? There are a number of his buildings that can be placed in this category, although they are very different in type and in the nature of the construction actually used.

The extensions to the Boots factory at Beeston and Odhams printing works at Watford were designed and built at much the same time. Both have long-span roofs but the planning requirements of the buildings led to quite different solutions.

Empire Pool

Williams' earlier involvement with the Empire Exhibition would have made him a natural choice as designer for the Empire Pool (now called the Wembley Arena) which was to be built on part of the site. He had already undertaken work for the client, Wembley Stadium Ltd at the 1924 Exhibition and was himself a shareholder. Indeed, the building was to be on the site of the ornamental lake, causing the *Architect and Building News* to report that:

> 'the ornamental lake…is to be deepened and covered in to accommodate international and championship swimming and to give seating for more than 8000 spectators.'[152]

Although the building was to house one of

Fig. 7.2. Empire Pool — exterior.

the largest swimming pools in the world, 200 ft x 60 ft, and also to be convertible into an ice rink and tournament arena, the reported seating capacity was an exaggeration. In the event it was to have sufficient raked seating to accommodate only about 4000 spectators but still a large number. These basic functional requirements called for a simple rectangular space with a large clear span to avoid any visual obstruction for spectators. In a lecture that Williams gave to the Architectural Association shortly after the building's completion he noted that the most difficult part of the design process was establishing the starting point. We have seen that for his bridge designs this became the overall economy of the structure; here at Wembley it came from the Middlesex County Council's Building Regulations. These determined the rake of the terracing for the spectators, the steps to have a going of 11 in. and a rise of 6 in. This, together with the fixed size of the pool, gave the overall size of the pool arena, and the horizontal grids for the structure. These were then the dominant features of the building's plan and interior (*Fig. 7.1a–b*). This determined the overall form of the building, although hardly the dramatic external expression of its structure (*Fig. 7.2*), especially what appear to be massive counterweights or fins along the sides of the building (*Fig. 7.3*).

Naturally, additional accommodation was also required. In the basement were the

Fig. 7.1a. General view of the Empire Pool.

Fig. 7.1b. Empire Pool — interior.

Chapter Seven
Long-span buildings

Fig. 7.3. View down the side of the Empire Pool with the dramatic structural fins.

changing facilities and the plant room. Along the western end of the pool Williams provided a two-storey structure to accommodate the main entrance at ground floor level and offices above. At the other end a similar two-storey extension provided for kitchens at the sides with offices above and additional poolside space on the ground floor behind the diving board. Unfortunately, the overall massing resulting from these extensions was not particularly satisfactory. The structures at either end of the pool were so small in relation to the pool itself that they have something of the appearance of later extensions, particularly as they had simple reinforced concrete frame structures, quite separate from the structure of the pool itself, and hardly relate well to the treatment of the gable wall of the main volume.

The early design for the pool was first published in the prospectus of the company (*Fig. 7.4*). The structure, which was to provide the main visual impression of the building, comprised a series of three-pinned in-situ concrete arches with an unprecedented span of 236 ft. These were to be above the roof that they supported and so clearly expressed. But what made the design so dramatic was that they were to sweep round at their ends in a semicircle to be continuous with the inward cantilevering balcony supports (*Fig. 7.5*). Inside the main space there was little evidence of any of this structure. The arches are above the roof and the cantilevers beneath the terraces. The only part of the interior where one has any sense of the structure is in the lower circulation areas where stair structures compete with the terrace supports (*Fig. 7.6*).

Perlmutter and Mark have noted that Williams adopted a planning module based on the space required for individual seating with a unit of 33 in. Based on this, the structural bay width is 22 ft, i.e., eight units.[153] Staircases, which give access to the terraces, are at 44 ft intervals. Between them and set between the structural frames are toilets, buffets and fire exit areas.

The practicalities of construction, in particular, the need for simpler shuttering, would have led Williams to adopt the simpler, if less visually satisfactory, rectangular fins that were actually built. It is perhaps the rather heavy appearance of these that have led to their being thought of as counterweights, although their real function is far simpler. Whether or not Williams thought of the original semicircular form as counterweights is impossible to say. Perlmutter and Mark have considered the present structure as acting in this way and have shown that the fins have a negligible effect on the horizontal force of the arch; their omission only reducing it by 10%. Of course, their real function is to carry the bending moments from the roof beam above, as is clearly seen from the drawing of the reinforcement (*Fig. 7.7*). Had the semicircular form been used it would have needed to be the same section throughout,

Fig. 7.4. Perspective of the Empire Pool prepared for the prospectus.

Fig. 7.5. Empire Pool — bird's eye view of the proposed building.

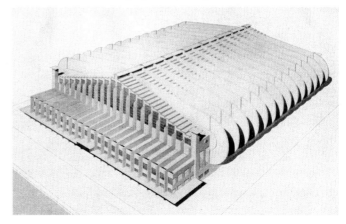

111

Chapter Seven
Long-span buildings

Fig. 7.6. Empire Pool — interior under the seating balcony showing the staircase structures.

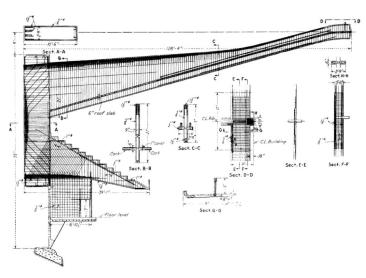

Fig. 7.7. Empire Pool — reinforcing drawing for the arch structure.

whereas the rectangular fins actually used were made wider than the beam itself in order to reduce the outward projection that would otherwise have been needed.

In other respects the final form is fairly close to the original proposal, although the extension behind the diving pool does not seem to have been anticipated. The original interior view shows glazing on a plain wall following the line of the roof and the balcony. The outward projection of the extension at this end, matching that at the front, means that this glazing is interrupted by the horizontal beams framing the extension roof. The awkwardness of these extensions and their conflict of scale with the remainder of the building is only exacerbated by their structural arrangement because, in keeping with his principles, Williams provided these annexes with a simple and separate frame structure instead of continuing the main structural form over the entire plan area and arranging the subsidiary space within it. Moreover, the secondary spaces have simple reinforced concrete frames at right angles to the main frames over the pool clearly articulated by columns formed as projecting fins. The other rather awkward feature of the building is the way in which the water towers are placed at the ends of the gables.

Odhams printing works (1935–38)

Williams was commissioned to design Odham's new printing works when they acquired new presses and relocated in 1934 to a site beside what is now known as the Watford bypass. Three years later, a second phase of the works was built but the two phases have very different structures although enclosed within the same envelope. These commissions are the kind of work that would have gone to an engineer, a simple factory and office complex that might have been carried out by a number of firms, but in Williams' hands the result was unusual, especially the second phase of the building.[154]

The printing works is rather conventional in its overall appearance and is certainly not an example of concrete architecture. Instead, it adopts the conventional arrangement of a long, two-storey brick office block dressed out with concrete string courses in the manner of the time, both fronting and forming the sides of a large flat-roofed shed (*Fig. 7.8*). Because brick walls return down the sides, the only place where the roof structure is visible is in the cantilevered loading dock at the back. However, the structure behind this conventional façade was anything but the norm for the time, which led *Architect and Building News* to comment that:

'In all his work, while efficiency for the purpose and common sense use of materials are the keynotes, there is always that touch of imagination that prevents it from being just dumb engineering — it bears the imprint of great personality.'[155]

For the first phase, Williams, rather unusually, employed a steel roof, although this is exactly what might have been expected for a building of this type. Deep steel trusses span 96 ft at 48 ft centres but also cantilever 48 ft at the rear of the building to provide a covered loading dock. These carry secondary girders at 12 ft centres. With the main spans 11 ft deep and

Fig. 7.8. Perspective of Odhams printing works.

Chapter Seven
Long-span buildings

the secondaries 4 ft deep, the overall depth of the roof structure is 15 ft. A feature of this that should be noted is that the webs of the upper plate girders were pierced to accommodate ventilation outlets. Ventilation was an important feature of the design because of the volatile materials used in the printing process and the scheme incorporated a separate solvent recovery building.

Additional headroom was required for the second phase so that to maintain the same roof line Williams had to reduce the overall depth of the structure. At the same time, in order to keep to the same grid, he had to change the structural arrangement. This was done by changing from a steel truss roof to reinforced concrete portal frames at the same 48 ft centres and 96 ft span as the steel trusses. The beams of the portal frame were 10 ft deep with 4 ft 6 in. deep secondary beams cast integrally with them (Fig. 7.9a–b). This arrangement gave the additional 5 ft of headroom required. The concrete structure also has the advantage that it does not require the regular painting that is needed for the steel structure of the first phase. When constructed this was the longest structure of its kind in Britain, although it received little attention in the architectural press of the day.

Boots Drys

In spite of the original intention that the Boots Drys building should be produced as a mirror image of the Wets, in the event it was built as a completely separate and quite different building (Fig. 7.10a). An early perspective produced by Williams' office shows the relationship between the two buildings and the general form that the Drys building was to take (Fig. 7.10b). A central block has extensive single storey areas either side and the railway lines serving the loading and unloading docks along the edges of the building are clearly visible. Given the close relationship between the form and the production processes of the original plan, we must assume that the changes were a result of a reappraisal of the production process. This may have been because of the reversion of the company to British ownership in 1932 or perhaps from experience of the operation of the first building. Although the building contained a number of striking structural features it did not have the clear simple

Fig. 7.9a. Sections of the Odhams printing works, first phase.

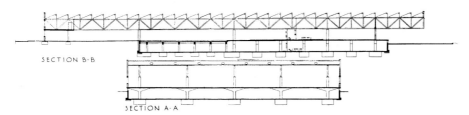

Fig. 7.9b. Concrete structure of the second phase of the Odhams printing works.

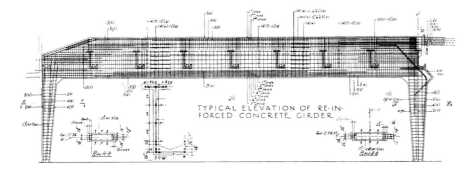

appearance of the first building and, perhaps as a result, did not receive the same attention from the architectural press. Nevertheless, it is a clear demonstration of Owen Williams' use of structure to accommodate functional requirements.

In this new building the elements of production were arranged vertically, rather than horizontally, in a five-storey spine block. The single-storey structures either side are, on the west, the unloading dock and sorting, and, on the east, packing and dispatch (*Fig. 7.11*). Compared with the Wets building, both the organisational arrangement and the resulting building form were turned inside out. The Wets building appears as a unified block with spaces carved out of it but with the most important distinction of function only seen inside in the vast atrium of the packing hall. In the Drys building the three principal areas are clearly expressed both in the external form and in the way in which their structures are handled. Even the main entrance to the new building, which was a simple recess in the early perspective, eventually came to project beyond the main bulk of the building producing a distinctive feature. This can be seen in the section (*Fig. 7.12*) and is in marked contrast to the handling of the entrance to the Wets.

The multi-storey structure was built using flat-slab construction with cruciform capitals and columns very similar to those

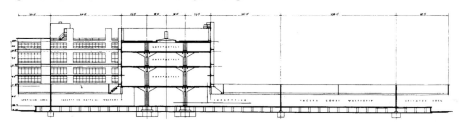

Fig. 7.10a. Cross-section of the Boots Drys building.

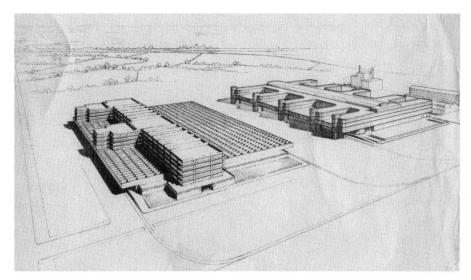

Fig. 7.10b. Perspective of the Boots site with the two buildings.

used in the Peckham building and in the Manchester Daily Express. The floors cantilever 24 ft beyond these columns, so that the structure here is very similar to the Manchester building. Two projecting wings of this long block also cantilever out over the unloading dock. For the single-storey elements, Williams designed an impressive

Chapter Seven
Long-span buildings

roof structure consisting of 9 ft deep Z beams cantilevering distances of 30 ft and 48 ft on either side (*Fig. 7.12*). In order to eliminate columns between the multi-storey production plant and the single-storey areas, the inside ends of the beams were carried on large concrete hangers suspended from deep beams across the roof (*Fig. 7.13*). The load on these hangers was thus ultimately carried by the two rows of columns that support the floors of the five-storey block. These hangers were clearly exposed on the elevations and owe their substantial bulk to the fact that they were made hollow to form the main extract ducts for the entire building.

In many respects, the structural design for this building was more closely suited to its function than was the design of the Wets factory. However, the external appearance of the buildings gives few clues to the ingenious structure. It can only be appreciated by looking at the cross section. The external treatment of the building does not conceal the structure, indeed it can be clearly read, but there are a number of competing effects. There are the clearly expressed hangers and ventilation ducts down the elevations of the tall block, vertical elements in an otherwise horizontal mass whose supporting function cannot easily be seen. The cantilevers of the single-storey blocks are dramatic but their effect is reduced because the Z-shaped beams are

Fig. 7.11. Boots Drys building — view of the unloading dock.

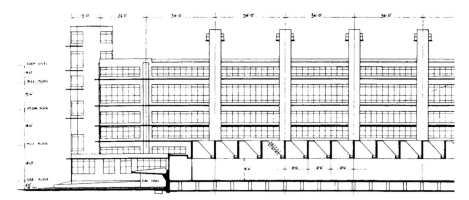

Fig. 7.12. Section of Boots Drys showing the structure of the packing hall.

116

Fig. 7.13. Boots Drys building, interior of the packing hall — the beam ends are suspended from hangers above the roof.

concealed behind a deep facia beam. Little of this would have been apparent from contemporary journal articles because of their rather scant coverage of this building.

BOAC hangar

The British Overseas Airways Corporation (BOAC) hangar complex is undoubtedly one of Williams' most dramatic building projects, its most spectacular features being the vast cantilever arch structures, which form the entrance to each hangar pen. But these visually dramatic features are not the only interesting aspect of the structure. Others, within the hangar complex, are concealed from above by the roof structure. Perhaps for this reason they have not attracted the kind of attention that has been given to the much simpler and thus much more dramatic forms of the hangars built by Nervi for the Italian airforce or by Ricardo Morandi for Rome airport. At the same time, Williams also designed the wing hangars that do have a simple form, smaller but far more elegant than the main hangars. But these too have been neglected by critics.

Williams' interest in aircraft hangar design dated from before the end of the Second World War, when he put a great deal of effort into the research and development of possible methods of their construction. This speculative work was because Williams realised that there might be a considerable potential market for civil aircraft hangars following the end of the war. He produced an innovative design that involved clustering hexagonal concrete roof structures into a variety of plans (*Fig. 7.14*). Models of these were sent to the Ministry of Civil Aviation some time before 1950. The ministry then invited Williams to develop his ideas further with a specific commission to build a new hangar, maintenance and office complex for BOAC. This was far more than just a simple group of hangars and although the Ministry of

Chapter Seven
Long-span buildings

Fig. 7.14. Proposal for aircraft hangars based on hexagonal structures.

Fig. 7.15. Bird's eye perspective of the BOAC maintenance headquarters.

Civil Aviation was the official client for this, BOAC was involved in the design from its inception. The result was the construction of a complex of hangars, workshops and offices.

The associated wing hangars were planned during construction of the BOAC Maintenance Headquarters. The Ministry of Civil Aviation decided to provide complementary hangar space for less complex maintenance operations. It was thought unnecessary to provide a complete built enclosure for these and so a design was proposed that would allow the main body of aircraft to be protected from the weather while leaving the tail projecting beyond the enclosure. An elementary plan and brief specification were put out to tender on a 'design and build' basis. Williams and W. & C. French, the firm already building the Maintenance Headquarters, collaborated to produce a design and price that was successful in a limited competition.

The plan of the main hangars
Although Williams' hexagonal system was seriously considered it was eventually discarded. While it enabled a large number of hangars to be provided within a limited perimeter, its geometry was restrictive and it did not provide the flexibility of space within to enable it to accommodate the subsidiary spaces that were required, i.e., the workshop and office space. Eventually, a 'pen' arrangement was adopted as the most

Fig. 7.16. Internal perspective showing the engineering hall flanked by hangar spaces.

flexible and efficient plan form, best appreciated from an early bird's eye view of the project (Fig. 7.15). Four pens were to be provided; two each side of a rectangular envelope, and separated internally by a large cross-shaped plan that provided the necessary workshop spaces and office accommodation. Each hangar pen was designed to accommodate two of BOAC's largest aircraft, which fixed the size of each pen at 336 ft x 140 ft.

The cross-plan form, which links the four hangars, contains a single-storey engineering hall along the longest arm (Fig. 7.16), although the structural design of these was to be considerably changed from the simple early idea shown in this perspective. The shorter arms have ground floor store facilities with multi-storey office accommodation above. The overall plan dimensions of the engineering hall are 867 ft by 140 ft but the principal workshop area is restricted to a 90 ft wide central space. The remaining strips of floor area on either side provide subsidiary spaces and entrance doors into the four hangars. The structure over the principal workshop area comprises a series of impressive concrete frames at 18 ft centres made up of raked columns either side supporting concrete arches of 76 ft spans one level below the roof structure. Next to the hangars, a combination of raked and vertical columns support three levels of galleries surrounding the workshop area (Fig. 7.17).

The office accommodation within the shorter arm of the cross varies from five to six storeys high. Like the remainder of the building, it was planned on an 18 ft grid but

Fig. 7.17. Interior of the BOAC engineering hall.

Chapter Seven
Long-span buildings

Fig. 7.18. Model of the completed BOAC maintenance headquarters.

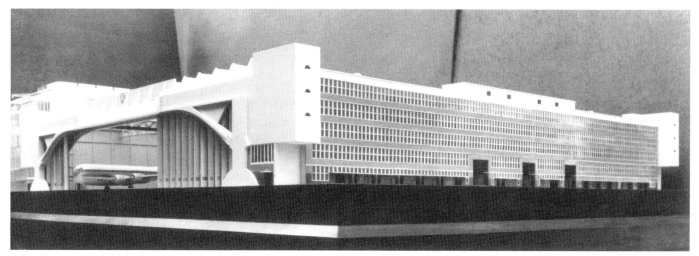

Fig. 7.19. Perspective drawing of the proposed BOAC maintenance headquarters with an exposed concrete frame structure for the offices.

its structure uses a conventional post and beam concrete frame. Williams also provided further office space in flanking buildings across the side walls of the hangars. Their elevations, in which a repetitive rhythm of windows is separated horizontally by long bands of brickwork, are the most disappointing aspects of the scheme because they give no clue to the ingenious structure behind. The relation of these offices to the hangar structure can be best appreciated from the model of the completed scheme (*Fig. 7.18*) but the early perspectives show that Williams originally envisaged a scheme that was more expressive of the structure of these offices (*Fig. 7.19*).

The same roof structure is carried over all of these spaces, having a total length of 420 ft. In roofing two hangar pens and the engineering hall between them it has three roughly equal spans of 140 ft. Over each of the hangar pens it is supported on the concrete frame structure of the workshops at the rear and the dramatic entrance structures at the front and comprises a series of 10 ft deep V-shaped trussed beams, constructed from both precast and in-situ concrete members. Each of the sloping faces of these beams is clad with patent glazing to provide uniform lighting within the hangars, with their lower connecting members designed to act as gutters.

The entrance door structures

The four entrance doors to the hangars are the only part of the structure that can be easily appreciated but these are dramatic indeed. They were built of in-situ reinforced concrete and have the unprecedented span of 336 ft. In essence they were designed and constructed like independent bridge structures, each comprising two cantilevers and an infill central beam. Each of these comprises a pair of structures providing a space between for the folding doors at the entrance to each pen (*Fig. 7.20*). When fully retracted the doors are located between the paired supporting pylons. These pylons are the cantilevers of these bridge-like structures; cast with a 90 ft projection complete with arched bracket support. Behind the supports of these are the huge counterbalances, each of which weighs about 1000 tons. This is most clearly appreciated from the construction photographs (*Fig. 7.21*). The shuttering for them was filled with mass concrete in a step by step process as the construction of the cantilevers, infill beams and the roof structure that they supported was carried out. In this way they balanced the ever-increasing loading imposed during

Fig. 7.20. External view of BOAC hangars.

Chapter Seven
Long-span buildings

Fig. 7.21. Construction photograph of the door structure and counterweights.

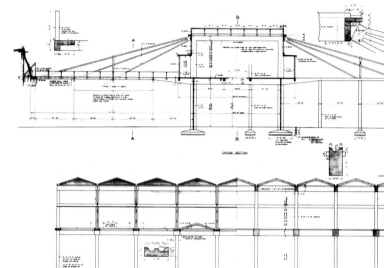

Fig. 7.22. Wing hangar sections.

construction. The central span between these cantilevers was then infilled with a complex 9 ft deep beam, the lower half of which had an inverted V cross section to bridge the space provided for the folding doors.

Wing hangars

The building comprised two hangars each 110 ft in depth and 565 ft long arranged back to back with a central engineering workshop between. This workshop, 60 ft wide and extending the full length of the hangars has a very simple two-story concrete frame structure from which the roofs of the hangars are suspended (*Fig. 7.22*). Light wells in the first floor admit daylight to ground level.

It was the roof structure that was the most notable aspects of this scheme. A series of 110 ft long members either side of the engineering workshop were supported at three points by raking ties of mild steel bars, which were embedded in granite for weather protection. These ties were simply welded to steel plates at their ends. Between the cantilevers there are slender precast concrete trusses located on a series of ribs cast into the cantilevers. These trusses in turn supported precast concrete purlins that carried a patent glazed roof.

It is difficult to know how to view these hangar buildings. The main BOAC building might well be regarded as one of Williams' greatest achievements. In planning terms the solution he produced to reconcile the complex organisation of almost one million square feet is masterly. Its structure is also unquestionably a masterpiece of reinforced concrete engineering, possessing a structural visual conviction unsurpassed by any comparable British building. Had Williams produced nothing else in his career, this building might warrant him the title of one of Britain's greatest designers of the twentieth century. Nevertheless, structurally it seems backward looking and Allan Harris has gone as far as to describe it as a folly.[156]

The reason for this rather harsh view of this structure is because engineering had moved on and Williams seems to have failed to move with the times. Freyssinet's pre-war invention of prestressed concrete had been adopted by British engineers and was being used in a variety of buildings, its use stimulated by post-war shortages of steel. But Williams never used this method of construction. The engineer who regarded mass concrete arches as more satisfactory than the reinforced concrete he had used in such a variety of ways was suspicious of this new technique and so failed to explore its potential for light, long-spanning concrete structures. Nevertheless, to ensure the structure of the hangars was an economical solution, Williams' reinforced concrete design was put out to tender in competition with a steel frame design produced by Ministry of Civil Aviation engineers. In this limited competition, the reinforced concrete design proved the cheaper (with the lowest tender submitted by W. & C. French Ltd). Of course, one has to wonder whether this was a fair competition.

Chapter Eight
Minor buildings

Chapter Eight
Minor buildings

This has not been a complete catalogue of Owen Williams' architectural work; that was not the purpose of the book because his influence has been through his major works rather than through the whole of his *oeuvre*. Nevertheless, an assessment of his own architectural work would hardly be complete without considering what might be regarded as his minor buildings. This reveals certain weaknesses in his skills as an architect, doubtless because the planning and form of these smaller buildings were not dominated by the overriding functional or constructional constraints that guided his designs for the larger buildings. He had no architectural training to fall back upon when this was an insufficient guide and the result was a certain clumsiness in his buildings, seen, for example, in the Lilley and Skinner premises and in the fire station at the Boots factory. He was more successful in the straightforward laboratory buildings at Pitstone and Thurrock.

Thurrock laboratories

The laboratory at Thurrock for Tunnel Cement was a small early building commission. This small-scale building, completed in 1934, seems to have caused Williams some initial difficulty because his early sketches are for a flat-slab scheme with four internal columns to support the roof

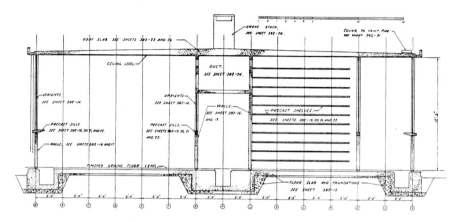

Fig. 8.1a. Thurrock laboratory cross-section.

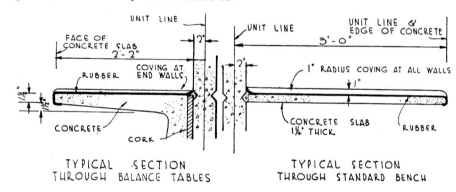

Fig. 8.1c. Thurock laboratory — concrete benches.

deck, hardly appropriate for a building of this scale. As the plan of this building was a simple double-loaded corridor, the simplest arrangement was to use the walls as structure (*Fig. 8.1a*) and he eventually devised a scheme that used the precast glazing mullions as structural support for the roof (*Fig. 8.1b*). These mullions were at 3 ft centres, each 4 in. square and 12 ft 3 in. high with a 3 ft 9 in. high in-situ wall between them and the glazing above. They were cast with grooves to take the glazing and the brackets that supported the work surfaces and the shelving. These mullions were not all that was precast because Williams also designed furniture and fittings of precast concrete: workbenches, shelving, bookcases, desks and fumigation cupboards — all of concrete (*Fig. 8.1c*). This almost suggests an obsession with the material but it was almost certainly the nature of the client's business that led to this apparent excess; a demonstration of the versatility of the product.

Stanmore flats

Whether or not this scheme had revived an interest in the possibilities of precast concrete it is impossible to say but immediately after the design of the Thurrock laboratories Williams set up a company to build standard apartments. Three blocks of four-storey apartments were built by this company at Stanmore in Middlesex, with the clear intention that they should form the prototype for other low-cost housing schemes using the same system (*Fig. 8.2*). As with the Thurrock laboratory, the basis of the structure was that the mullions were the principal structural elements. However, in this instance they were cast in-situ. Each was 15 in. by 5.5 in. at 3 ft intervals. Where glazing was not required the space between was filled with cavity brickwork, which was rendered externally. At 7 in. the cavity was unusually wide but this was to provide space for services. The mullions carried

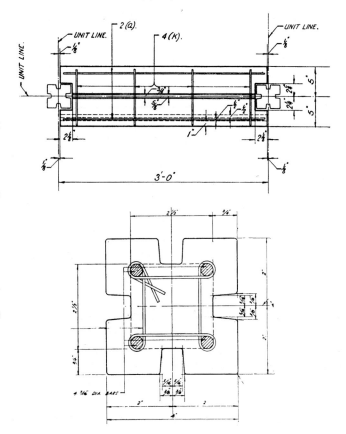

Fig. 8.1b. Thurrock laboratory — published construction details.

Chapter Eight
Minor buildings

Fig. 8.2. Stanmore flats.

concrete crossbeams, which spanned the full depth of the building and in turn carried timber decking for the floors and roof.

Williams may have been thinking of precast concrete while deciding to build using in-situ concrete and brickwork for these trial flats. They have a little of the form adopted by Auguste Perret in France who used expressed vertical elements with precast concrete panels between them and one wonders whether Williams had contemplated using precast concrete rather than rendered brickwork for at least the external leaf of the wall. The construction was no doubt ingenious and carefully considered. The internal leaf, for example, was of bricks on edge, presumably regarded as a cost saving on normal brickwork. The problem that Williams was facing was that he had little experience of domestic construction and Harris recalls that architects regarded the result as risible.[157]

Lilley and Skinner

Williams repeated some of the devices that he had used for the Sainsbury's warehouse in his much later building for Lilley and Skinner. The building had to be on a constrained L-shaped corner site, the accommodation comprising seven floors of shoe storage with three floors of offices over that. Williams used the same, boldly expressed staircase tower and the flat-slab construction of Sainsbury's building but with edge columns clearly expressed on the outside. These were square-section columns with projecting brackets rather than mushroom heads. Only the free-standing column at the turn of the L and its corresponding side wall column had mushroom heads of the same square form as used in the Wets building. The top three floors stepped back over part of the front with the top two floors stepping back over the remainder and this produced a slightly awkward massing which Reilly described as *'clumsy'*, deciding that the building had *'little to commend it externally'*.[158]

The *Architect and Building News*, however, was prepared to admire it as *'a well balanced solution of a difficult problem'*.[159] This journal was always ready to admire the straightforward functional building. But a feature that neither commentator remarked upon was that the boldly expressed columns of the warehouse floors disappear at the office level where the glazing became continuous. This is because Williams simply raked back the columns to carry the stepped-back floors. In doing this he was using a device that had previously been used extensively in London by designers in structural steel to provide for the set back of attic floors. Here, however, the rake on the columns was somewhat greater, which would have resulted in greater horizontal forces.

Dollis Hill synagogue

At Dollis Hill synagogue the problem was simply to provide a central space for worship with subsidiary spaces at either end. What is striking about the design is that Williams used a folded plate arrangement for both walls and roof, which was certainly an unusual structure and possibly unique for the time; folded plates were a largely post-war phenomenon. Forming the relatively thin walls in this way would have stabilised them but a thickening at the folds also enabled the walls to carry the cantilevered galleries. Like the Pioneer Health Centre, there are two perspective drawings for this building showing the early design ideas (*Figs 8.3 and 8.4*). The overall massing of both designs is substantially the same and similar to the Empire Pool, in that Williams used a tall long-spanning structure over the main space but lower structures over the subsidiary spaces at either end. However, both perspectives differ from the final building in the relative volumes of the subsidiary spaces. In the perspectives, the volumes of the subsidiary blocks are similar in size whereas planning requirements have resulted in a large block at the front, four bays deep rather than three, with a corresponding reduction of the depth of the block at the rear. Where the perspective drawings differ is in the handling of the junction between the large space and the

Chapter Eight
Minor buildings

Fig. 8.3. Perspective design for Dollis Hill synagogue.

Fig. 8.4. Alternative design for Dollis Hill synagogue, not as built.

subsidiary blocks. The unbuilt scheme tends to emphasise the folded-plate nature of the walls with the main space more or less the same width as the subsidiary spaces. However, the need for thickening of the walls at the folds meant that the end wall planes were longer than the others. The sloping roof also required an overhang at the eaves, which gives a strangely domestic touch to the building.

The adopted design abandons the clear articulation of the structure and adds what must be an unnecessary thickening of the walls at the corners of the main space. This makes the central block read as a rectangular box from which the folded walls protrude and rather less like a folded paper form writ large in concrete. The result is three rectangular blocks, with the central one both slightly wider as well as higher than the other two (*Fig. 8.5*). This is certainly a better composition but the nature of the structure is not so clearly expressed. According to Williams, the windows within the walls were positioned and shaped according to the forces. The walls are vertical girders and he claimed that the windows within the webs of these girders were shaped according to the diagonal lines of shear stress, it merely being convenient or coincidental that they thus accommodated the symbolic Star of David. Of course, he could hardly use any such structural argument to justify the other

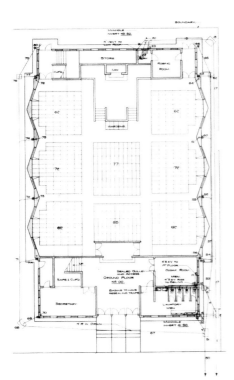

Fig. 8.5. Dollis Hill synagogue — plan.

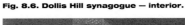

windows conveniently shaped to represent the form of a menorah.

Internally, the building is as plain as it is externally. Almost the only materials used are the concrete of the structure and the light timber of the furniture. Apart from these, simple metal guard rails are provided at the steps at the front and on the upstand concrete balcony rails where steps come down to them (*Fig. 8.6*). The stepping balcony is carried by ribs that project from the folded wall, as seen from the reinforcing drawing (*Fig. 8.7*). This also shows the small windows that were incorporated into the walls just below the line of the verge.

The building was not a great success and had mixed reviews. The correspondent of the *Architects Journal* suggested that Williams was trying a little too hard to be an architect. But apart from the overall form the finishes of the building were poor. The plywood shuttering bowed so that the walls were not

Chapter Eight
Minor buildings

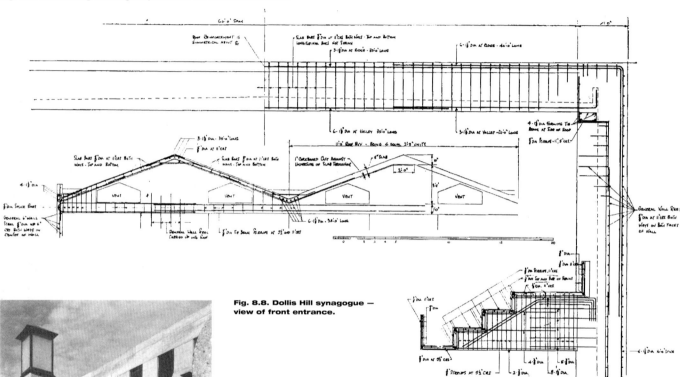

Fig. 8.7. Reinforcing drawings reproduced in *Architects Journal*.

Fig. 8.8. Dollis Hill synagogue — view of front entrance.

perfectly flat, and they leaked at the joints, leaving a rather prominent grid over the surface (*Fig. 8.8*). Moreover, early photographs suggest that there was some colour variation between lifts. The clients were also unhappy with the result and for a time legal action was threatened.

Fire station

The fire station at Boots was one of a number of subsidiary buildings on the site; a tiny building and part of the later extensions there. One might have ignored it along with many other completely utilitarian buildings that Williams put up as adjuncts to his factory buildings were it not that it shows a quite self-conscious attempt to produce something more than a mere utilitarian building. It consists of a series of concrete frames with the walls and roof set in their inside face so that the structure is clearly expressed (*Fig. 8.9*), in some ways a minor version of the Wembley Arena. The building housed two fire engines but with lower rooms on either side of the main garage space to provide for a rest area for the firemen and an equipment store. The roof over these lower spaces was cantilevered from the main columns, also with the beams above the roof. This appears to be grossly over structured and the only purpose of cantilevers seems to be that they allowed these smaller spaces to have three sides of glazing uninterrupted by any columns. At the back of the building is an elaborated hose-drying tower. Square on plan it comprises four concrete lattices infilled with glass blocks.

If the correspondent of the *Architect's Journal* thought that Williams was trying a little too hard to be an architect at Dollis Hill, then this is surely another example of the same. Something rather more straightforward would have sufficed and the whole composition, which is scarcely beautiful, gives the impression that Owen Williams was having fun.

Fig. 8.9. Boots factory fire station.

Provincial Newspaper

A far more successful example of planning on a difficult site than the Lilley and Skinner building was that for the Provincial Newspaper offices. The only journal to cover this was the *Architect and Building News* and it made a point of explaining the clever planning involved.[160] The site was extremely limited, only 30 ft x 35 ft and the accommodation was arranged over 10 floors (*Fig. 8.10*). Circulation space was kept to a

Chapter Eight
Minor buildings

Fig. 8.11. Provincial Newspaper offices — floor plan.

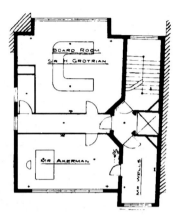

Fig. 8.10. Provincial Newspaper offices.

minimum by placing the staircase behind the lift shaft and by splaying the walls to provide a lobby for the latter (*Fig. 8.11*). Lavatory accommodation was in the projecting part of the building over the front entrance. The structure was simple and based on load-bearing concrete walls that were cast in two lifts to each floor. The projecting string-courses, one midway up each floor, the others forming a drip moulding over the window heads, were also there to enable the shutters to be levelled at each lift. The journal commented on the high level of day-lighting within the offices but the structural problem presented by the large window openings was that these walls then provided little resistance to wind loading on the side elevation. To cope with this the side wall and the strips of front and back walls at the corners were together treated as a channel section in bending.

Pitstone laboratories

When Williams came to build laboratories at Pitstone (1937–39) for the same client as his earlier cement laboratories, he was presented with a rather different problem. The requirement for a canteen, offices and reception area led to a two-storey building with a two-storey high entrance and stair. The frame was still clearly expressed on the external walls but now large windows were shaded by deep (4 ft) projections of the floor and roof slab only interrupted by the glass-block wall lighting the entrance lobby (*Fig. 8.12*). The logic of these projections may have been to provide the necessary shading but architectural consistency seems to have required that they be carried right round the building. One cannot imagine they would be of any use other than on the south facing elevation. It is clear from this that Williams was exercising a little more than strict functionalism in the development of his designs.

Newspaper buildings

Apart from the BOAC hangars, there are two post-war architectural projects that need to be mentioned, both for newspapers. The first was the Daily Mirror building near Holborn Circus in London designed in association with the architects Anderson, Forster and Wilcox, but, although it was one of the most complex building projects that he worked on, it had little influence on the architecture of the time.

Fig. 8.12. Pitstone laboratories showing the effect of the sun screens.

Chapter Eight
Minor buildings

Completed in 1961, Williams used an ingenious method of construction to save time in the basement. The scale of the structure with 45 ft spans was, like his other newspaper buildings, determined by the press runs. The basement columns were first excavated and cast, and then the ground floor slab. This enabled the construction of the superstructure above to proceed at the same time as the excavation of the basement. Once the latter was completed the basement columns were underpinned with concrete rafts. Both the planning and structure of the building were complex. For example, within the basement structure was hung a complete floor supported by hangers on a 15 ft grid. The structure of the tower block above the third floor is also based upon a 15 ft square grid. To carry this smaller grid there are a series of arches within the floor to ceiling height of the third floor to transfer the loads from the floors above to the 45 ft grid of columns below.

Fig. 8.13. Proposed extension to the Manchester Daily Express building.

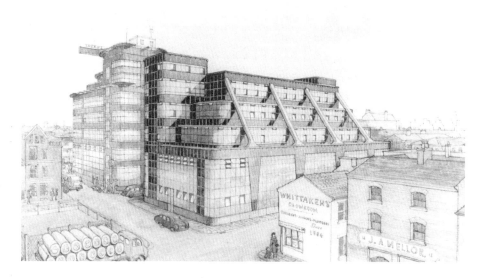

This complex design makes this one of Williams' best building projects when judged from an engineering point of view. However, this did not result in any interesting architecture. It appears simply as a twelve-storey tower surrounded by subsidiary accommodation of varying heights up to four floors. With solid end walls and simple fenestration, the building is quite undistinguished compared with much of the commercial architecture of the period. Perhaps there is a parallel to be drawn here with his work at the Fleet Street Daily Express building, where his ingenious solution to the problem of existing steel columns in the basement went unnoticed at the time. The engineering may make a significant contribution to an architectural project but it may be handled in such a way that it is completely invisible.

The other building was an extension to the Daily Express in Manchester, rather hidden away at the back of the original building but substantial in its own right. Only the first stage was built with tapering columns clad in black to match the original building. Provision was made in the design for the building to be extended upward with set backs at each floor to meet the planning requirements. The intention was to use raking columns for this, much as he had done at the Lilley and Skinners building, except that here they were to be exposed. A perspective produced by Williams' office shows what was intended (*Fig. 8.13*).

Williams as an architect

Set against his better publicised work the minor buildings enable us to form an overall view of Williams' ability as an architect. In this it is clear that he had his limitations and we might be tempted to agree with Faber that engineers were not competent to deal with aesthetic issues. This Faber blamed on their education.[161] However, this is perhaps a rather harsh judgement. It is clear that Williams had no particular architectural intentions, no fixed architectural language that he was exploring and developing. Each building was a completely separate problem. The only buildings that show any consistency of treatment are the several buildings for the Daily Express that followed on from the Fleet Street building and this may be read as a house style for the newspaper.

Nevertheless, examining the buildings in detail shows Williams to be not simply a skilful structural designer but also someone who was capable of handling the planning of buildings and the integration of the services. The planning of the Dorchester and the Boots Wets building are sufficient to demonstrate the former, although it can also be seen in the small Provincial Newspaper building. Integration of services with the structure is most clearly revealed architecturally in the Boots Drys building, where he seems to have anticipated the work of some post-war architects, although perhaps with more functional justification and less for the stylistic effect of such an approach. It was also provided for by the multiple columns of the Dorchester that facilitated pipe runs within and by the handling of the ventilation extracts at Odhams press. Even at a detailed level we see his concern for this aspect of the design in the careful inclusion of sleeves for pipes through his column heads.

The minor buildings were largely ignored by the architectural press of the day, the *Architect and Building News* being the only journal to cover them all. Of course, they were buildings of no great significance and we would hardly note them if they were the work of an architect that we were considering. The *Architect and Building News* seems to have taken an interest in Williams work because of its concern for the technicalities of building, its espousal of a functionalist approach and its interest in Modern Movement architecture. In this regard, Williams was not the only architect in which the journal took an interest. However, what is of more importance is the reception of his other buildings by the wider architectural press, and the influence these may have had on the development of Modern Movement architecture in Britain.

Chapter Nine
Assessment

Chapter Nine
Assessment

The contribution of many engineers to the architecture of the inter-war and immediate post-war years was to provide new structural forms in a period when both engineering techniques and architectural ideas were developing rapidly. For many this involved the inventive use of reinforced concrete, seen, for example, in the work of Owen Williams' contemporary Oscar Faber, or later Ove Arup who came to consultancy work fairly late in his career. For others it was innovative designs in structural steelwork as exemplified by the work of Samuely, the engineer to Emberton for Simpson's store in Piccadilly, and Gropius/Chermayeff for the de la Warr Pavilion. While these were, of course, very visible and well-publicised examples of Modern Movement buildings, there were equally important contributions made to the architecture of the time by other engineers whose structures were not so dramatic but whose abilities allowed architects to achieve their visionary ideas either in traditional or modern buildings.

Williams' contribution does not fit into either of these models. For one thing, the technologies he used were not in themselves particularly innovative. For example, flat slab construction, which he used to such effect, had been available and well known for many years. Indeed, one could argue that from a technological viewpoint he was a conservative, citing, as an example, his refusal to embrace the new techniques of prestressed concrete in the years following the Second World War. Unlike the majority of his engineering contemporaries, however, he practised as an architect in his own right and it is in this light that his main contribution should be judged. Having abandoned a collaborative approach towards his ambition for an architecture of concrete, he was determined to achieve this himself through the application of a functionalist philosophy. While it was common for engineers to design industrial buildings at that time, there could be no one with Williams' reputation nor with the good fortune to be given an early architectural commission in Park Lane. Within two years of his appointment as architect to the Dorchester Hotel he was involved in the delivery of three buildings, which were all regarded as quite spectacular and whose reputation has remained undiluted. One of these buildings, the Boots Wets building, made a very special impact and it is still seen as having seminal importance to the development of British architecture of the twentieth century. Quite apart from this very tangible evidence of his contribution, it is possible to observe from his career other, perhaps less obvious, contributions that Williams made to the architecture of our time.

Williams and concrete

The one unifying theme in Williams' career was his commitment to reinforced concrete, predominantly in-situ concrete, but occasionally precast. His reputation for this was first established in collaboration with Ayrton when they introduced the material to a sceptical British public and, indeed, a sceptical architectural profession, through the buildings of the Empire Exhibition at Wembley in 1924. Together they were able to demonstrate that concrete was highly adaptable and that it had serious architectural potential. The towers of the stadium remained a symbol of Wembley for the rest of the century and this work gained Williams a knighthood and a public reputation as Britain's foremost engineer in the material. For almost a decade he was affectionately referred to as 'Concrete Williams' in the British press.

The obvious question is why Williams should have persisted with concrete for the design of frame buildings when steel was an obvious alternative. His commitment to the material was just short of an obsession. He only rarely used structural steel and in one small project at Pitstone went to the extent of designing and manufacturing concrete furniture. This passion for the material, evident in all his projects, was also shown in published papers and presentations. In 1931, for example, he delivered a paper to the London Society entitled 'The Portent of Concrete'.[162] His aim in this paper was to demonstrate that on purely economic grounds concrete had clear advantages over its competitors, principally steel and concrete. As far as steel was concerned, he claimed that the cost of carrying the same compressive load on steel was three times that of carrying it on reinforced concrete, even without allowing for the additional cost of fireproofing the steel. For members in tension he claimed that the labour cost of fabricating steel was greater than that required for placing steel bars in reinforced concrete. The disadvantage of brickwork was its weight, which meant that it could not be built to the same height as concrete, and a brick structure seriously restricts the amount of daylight penetration. He suggested that the cost of a brickwork pier was almost ten times that of a structurally equivalent concrete pier. However, his conclusion that greater economies were possible if the height limits of the LCC were increased to a realistic figure of 500 ft, made possible by reinforced concrete, seems to have ignored the fact that these were based on fire-fighting requirements.

All this led to the general argument that as reinforced concrete was cheaper than its competitors, structural steelwork and loadbearing brickwork, it would eventually supersede both for structural purposes.

However there were obvious fallacies in this argument. Had he compared the height limits of reinforced concrete and steel, the latter would have been shown to be superior because of its lighter weight. At the time, Oscar Faber defended structural steelwork in a well-argued paper, exposing the weaknesses in Williams' calculations.[163] Using the LCC regulations and detailed costings he showed that a steel stanchion cost no more than 1.71 times the cost of a comparable concrete column, roughly half the difference claimed by Williams (1.71 times for an encased stanchion and 1.5 times for an uncased one). However, based on his considerable experience in designing all types of building he pointed out that even the framework of a minimal factory building only represented about 10% of the overall costs; for a more pretentious bank building the figure was 6.6%. Thus, the overall cost saving of a concrete frame rather than a steel one was in the region of 3.5%, a figure only based on a comparison of compressive loads; if tension conditions were compared the figure would more realistically be about 1.6%. To be set against this minimal cost advantage of reinforced concrete were the many advantages of steelwork: reduced size of structural members, thus improving floor to structure ratios, and speed of construction. Faber wrote:

'I should not have dealt with the matter so

Chapter Nine
Assessment

extensively if Sir Owen had not stated that the onus of proof lies on the structural steel advocate. May I hasten to add that while Sir Owen appears to be an advocate of reinforced concrete under all circumstances, I have a strictly open mind on the question and consider that there are cases when either is the better material to adopt for a given work, and act accordingly in my practice; and I should not like to be considered as an advocate of either material for all purposes.'[164]

One year later, the influential *Architectural Review* published a more emotional justification for the world dominance of concrete in a short piece by Williams entitled 'A Concrete Thought'. In it he described the *'joy of contact with a universal material'*.[165] He opined that in a world reduced in size the universal use of a material common to all nations would encourage people to think internationally rather than nationally. Although this suggests a certain political naïveté to today's readers, in the context of the 1930s this type of propaganda was not uncommon, particularly in the field of modern architecture. In many ways it reflects similar simplistic statements emerging from protagonists for the Modern Movement in Britain.

It must be remembered that at this time the Modern Movement in Britain was at least ten years behind developments on the continent. Le Corbusier's book, *Vers une Architecture*, translated as 'Towards a New Architecture', although appearing in France in 1923 was not published in Britain until 1927. The book made a massive impact on the British avant-garde. Within its pages were strong arguments in favour of the engineer's approach to design and the clear conclusion that only through an application of this approach could a new architecture emerge. Not until 1933 did a formal grouping emerge in England to positively promote new architecture. This was the MARS (modern architectural research) group. It comprised a small group of mainly young architects and artists who for the remainder of the 1930s spearheaded arguments for the development of modernism in Britain. Opposing them was the influential conservative group of architects who argued for a traditional approach to design based on the *Beaux Arts*. The traditionalists saw no contradiction in clothing steel frame buildings with façades of stone styled in the classical tradition. In truth, the vast majority of built work within the 1930s fell between these extremes but nevertheless the debate itself was a dominant feature of the period.

Williams was never directly associated with British modernists and was never part of the MARS group. But to many in this group, particularly Wells Coates, he was something of an inspiration in the early years. Here was an engineer committed to the most modern of building materials — reinforced concrete — abandoning collaboration with architects and determined to apply the engineer's design method to develop an architecture of concrete, themes taken almost directly from Le Corbusier's book.

We have already looked at the way in which Owen Williams had developed his architectural ideas from his reading of Mendelsohn but that this was closely linked to his interest in concrete seems to have been in tune with the ideas of the Modernists. They too showed a bias in favour of this new material of the time. Bennet and Yerbury's book on concrete architecture showed how the architectural potentials of this material were just being exploited on the Continent. Examples in Yorke and Gibberd's books on modern flats and houses show a preponderance for the use of this material compared with the alternatives. Even the flats in England that they described tended to be in concrete, in spite of the steel interests' attempts to promote their material for flat building. Williams must surely have also read Trystan Edwards' attempt to address the aesthetic treatment of structures, such as bridges and factories, for which engineers might be responsible.[166]

Williams as an architect

The importance of the timing of Owen Williams' entry onto the architectural stage can be seen in the attitudes towards the Dorchester Hotel. At the time of Horder's resignation over Williams' design, the *Evening News* reported that:

> 'Other experts say that in the silent struggle between the modern and the ancient forces in society the new forces are bound to win in Park Lane.'[167]

These other 'experts' were presumably aficionados of the Modern Movement speaking more in hope than from any clear judgement of the situation. Certainly there were different forces at work within the architecture of the day and Williams was to some extent a player in this game. However, the 'new forces' were not then as strong as some seemed to believe. When Owen Williams himself resigned three weeks later, the *Daily Telegraph* was speculating that the job would go to Lutyens, hardly a 'new force' in architecture.[168] Neither could Curtis Green be ranked as such; the forces of conservatism were stronger than some hoped at the time.

The furore surrounding Williams' appointment to the Dorchester Hotel project, which generated extensive press coverage, epitomised for many the state of British architecture at this sensitive time. From the debacle, Williams attracted valuable publicity and it is not surprising that the small group of British modernists eagerly awaited his next commissions. They were not disappointed. Within twelve months the Daily Express made its entré and although Ellis and Clarke were credited along with Williams there was little doubt that its architecture had been driven by the dramatic engineering of Owen Williams. The article by Chermayeff in the influential *Architectural Review* was a defining moment. In this article real recognition of the contribution engineering could make to architecture and Williams' primary role in the conceptual design of this building was made abundantly clear.

We have already seen how Chermayeff and others failed to fully understand the extent of Williams' contribution, concentrating only on the clearly defined frame of the extension and ignoring the messy, though structurally very bold, adaptation of the original building. What he concentrated on was the effect of a bold structure on the planning and elevational treatment of the building. If the acceptance by Curtis Green, and both the architectural and more general press, that he was the architect of the Dorchester Hotel, while Williams' masterly planning was completely ignored, gives the impression that architecture was seen as something only skin deep, then this is reinforced by the reception of the Daily

Chapter Nine
Assessment

Express building. The Modern Movement was still young and there were few in Britain who were able to appreciate how structural inventiveness could contribute to the planning of buildings. When it came to the structure, the architectural community still thought largely in terms of simple repeated grids of columns with the resulting frames merely sized by specialist subcontractors rather than having the structure designed by a consulting engineer.

It was only with the building of Boots' pharmaceutical factory that the architectural community really took notice of Williams' architectural skills. This was no ordinary factory building. It was huge in its overall concept and still large in comparison to many other factory buildings of the day. The basic method of construction, the flat slab, had been used before in British factories, notably by Thomas Wallis who had demonstrated how it could be used to achieve a horizontal effect. This is what Owen Williams did at Beeston although going one step further by providing extensive glazing and a rather larger than usual mushroom heads to achieve a quite dramatic effect. This was surely no simple economic use of materials and the treatment that he adopted enabled the structure to be clearly read through the glazing.

As was to be expected, the journal that gave this building most space was the *Architect and Building News* with considerable coverage of its construction in the form of progress photographs. But other journals also noticed it. *Building*, with its bias towards the technology of construction, provided good details of the columns and the roof structure of the atrium. The *Architectural Review* made a particular point of the cantilevered roof over the loading dock, considering it the principal feature of the building. It was also noticed in the following year by Llewellyn Williams in his article on glass where he observed that:

'Mies van der Rohe dreams of a future city where the buildings will be a transparent membrane of glass stretched over the reinforced concrete skeleton, a condition which has been almost realized for the first time in this country in Messrs Boots Factory at Beeston designed by Sir Owen Williams.'[169]

The timing of the building was particularly opportune because in the meantime the *Architectural Review* had produced a special issue on steel and concrete construction.[170] In this the factory was illustrated under the heading 'mushroom slab construction' and described as having *'tiers of fungoid forms rising shadowily [sic] behind their glass screens'*. Presumably this was intended to be complimentary because Williams' building was here set alongside interiors of Alvar Aalto's 'Turun Sanomat' building at Abo, Finland, a newspaper building that also used flat-slab construction to dramatic effect in its interiors.

It was the Empire Pool building that was completed next and so was published next. There had been some anticipation of the project by *Architect and Building News* with publication of the prospectus drawings but far more significant was the wide coverage that the building had on its completion. Of course it was the dramatic structure of this that attracted the attention of engineering as well as architectural journals and if Owen Williams had intended to make a bold architectural statement that would get him noticed, he could hardly have done better. Nowhere before had a structure been so boldly expressed externally and the drawing of its reinforcement must have been the most widely reproduced drawing of the day, appearing with almost every account of the building. For some journals it was all they did publish of it.

We have already seen how the *Architectural Review* treated the Pioneer Health Centre under the pen of J. M. Richards. Like the Empire Pool, this was also a widely published building. Both attracted the attention of overseas journals, although in the latter case it was naturally because of its unique function. The most complete collection of photographs was produced by the *Architects Journal*, possibly the only journal to include a photograph of the main entrance to the building, and a view of the back alley. However, it had little critical comment on the building, although this was quite normal for the journal that generally restricted itself to simple factual accounts. This journal also seems to have become interested in the flat-slab construction of the floor because it later had a drawing of its reinforcement in its 'Working Details' series.[171] Perhaps the attention that had been given to the structure of the Empire Pool with the publication of the reinforcing drawing for the roof had raised architects' awareness of the fact that reinforced concrete could produce something quite different from the structures that they had been used to. It seems unlikely that many architects would have fully understood such details, and would certainly not be producing them themselves, but the same journal subsequently published reinforcing details for the structure of the Dollis Hill synagogue.

In their own terms each of these buildings was unique and worthy of special attention but it was undoubtedly the timing of their delivery that gave them particular significance as contributions to British architecture of the period. Had they arrived five years later it is likely that their impact would have been far less and their contribution to architectural development would have been greatly diminished.

This was certainly the case with Williams' later buildings. They received much less attention and are not generally regarded as having made the same contribution to developments in modern architecture. Part of this was no doubt due to the increase in examples of modern architecture but it may also be due in part to the changes in Williams' work itself. The smaller projects, which he produced later, were awkward and clumsy. But even his larger buildings from the mid-1930s showed distinct changes from the earlier clarity of the Boots Wets building and the Daily Express scheme in Fleet Street.

In many ways the later larger buildings exploited structural techniques in a more sophisticated manner but with less dramatic architectural effect, seen particularly in the Boots Drys building. This was surely as clear an expression of Williams' functionalist approach as his Wets building; perhaps even more so as less concessions were made to achieve a simple clear architectural statement through the use of structure. And yet probably for that very reason it received much less attention. There is a clear functional expression of the various elements: delivery and sorting, manufacture, packing and dispatch, and the expression of the ventilation ducts so ingeniously combined with hangers. In terms of structural engineering it is a masterly example of his work but the variations in technique used to respond to a range of

Chapter Nine
Assessment

different planning functions meant the building simply did not resolve into a simple and clear architectural statement. It was such a contrast to the Wets building that it failed to attract any significant comment in the architectural press. It seems that functionalism was a fine sounding creed but it was not enough by itself.

The Drys building, and perhaps Williams' smaller structures, say more about the relationship between structure and function than the Wets building or his other buildings where structure is dominant. In addressing the functional requirements of a building, structure can play a dominant or subservient role, even within the context of the functionalist modern architecture that Williams strove to create. Perhaps Williams came to appreciate this as his development as an architect progressed. Moreover, it would be impossible to pretend that Williams' designs were entirely functional. The development of his designs, as evidenced by the perspectives that were produced, demonstrates the exercise of clear architectural judgement.

Assessing the contribution of any designer through his work constantly varies over time, as we see history in a changing light. In architecture the timing of specific projects or ideas is crucial to their influence. This was undoubtedly so of Owen Williams' early architectural work which came at such a significant time in the development of modernism in Britain. But there is a longer perspective to consider from our present vantage point in the opening years of the twenty-first century, as we look back on the architectural developments of the remainder of the twentieth century. The much derided brutalism of the 1960s and 1970s, an unpopular period of British architectural history, is now itself being reviewed and reappraised and important projects are finding their rightful place in history. The frank, bold use of reinforced concrete, which characterised this work, has much in common with some of Williams' architectural work. Thus, as well as the influence that he had on his contemporaries we should perhaps now also consider his possible contribution to the foundations of British brutalism.

Endnotes

Preface

1. Cottam D. (1986). *Sir Owen Williams, 1890-1969*, with contributions by Stephen Rosenberg, Frank Newby and George Crabb, edited by Gavin Stamp. Architectural Association, London.

Chapter One

1. Bennet T. P. and Yerbury F. R. (1927). *Architectural Design in Concrete*. Benn, London.

2. The problems of the collaboration between architects and consulting engineers during this period that may have been a contributory factor has been discussed by Yeomans D. (2000) 'Collaborating with consulting engineers', in *Twentieth-Century Architecture and its Histories*, L. Campbell (ed.), Society of Architectural Historians of Great Britain, pp 125–51.

3. At Brixton School of Building, 'Editorial Notes', *Concrete and Constructional Engineering*, 6, 1911, p 651.

4. Marsh C. F. and Dunn W. (1906). *Reinforced Concrete*. 3rd ed. Constable, London. It was also Dunn that gave the first course of lectures.

5. 'Joint Reinforced Concrete Committee', *Journal of the Royal Institute of British Architects*, 13, 1905–06, pp 338–40.

6. See Yeomans D. (2000). 'Collaborating with consulting engineers', in *Twentieth-Century Architecture and its Histories*, L. Campbell (ed.), Society of Architectural Historians of Great Britain.

7. Kahn M. (1917). *The Design and Construction of Industrial Buildings*. Technical Journals, London.

8. *Architect and Building News*, 11 Sept. 1912, p 285. Designed by the architect A. Sykes. Williams was probably involved in this project although only in a junior capacity.

9. Williams' Papers, Birmingham. The photograph file includes a number of photographs of this structure and there are calculations also preserved.

10. Kahn M. (1917). *The Design and Construction of Industrial Buildings*. London. Plates 4, 59 and 60.

11. Williams O. 'The economic size of concrete ships.' *Engineering*, 107, (1919) pp 195–97.

12. Advertisements for the firm's products appeared in *Concrete and Constructional Engineering* during 1919 and 1920.

13. See *Concrete and Constructional Engineering*, 13, 1918, pp 503–12.

14. See White R .B. (1965). *Prefabrication: A history of its development in Great Britain*. HMSO, London.

15. Oscar Faber's book *Reinforced Concrete Design* went into several editions, beginning in 1912.

16. Job book of the firm, still held by Hurst, Pierce, Malcolme.

Chapter Two

17. Luckhurst K. W. (1951). *The story of Exhibitions*. Studio Publications, London.

18. *The Builder*, 128, 1925, p 226. Report on the address by Owen Williams to the Architectural Association on his work at Wembley.

19. Blomfield A. (1981). 'The use of Portland cement concrete as a building material.' *RIBA Sessional Papers*.

20. Collins P. (1959). *Concrete, The Vision of a New Architecture: A Study of Auguste Perret and his Predecessors*. Faber & Faber, London. Chapter six, p 112 *et seq*.

21. *The Builder*, 102, 1912, p 319.

22. 'The surface treatment of concrete buildings.' *The Builder*, 101, 1911, pp 179–80.

23. Black W. 'The architectural treatment of reinforced concrete.' *Kahncrete Engineering*, 1, 1914, p 33.

24. This lecture to the Concrete Institute, recorded in *Transactions of the Concrete Institute*, 3, 1911, p 246, is discussed by Collins (1959), pp 134–35, in some detail. (see note 20)

25. *Ferro Concrete*, Feb., 1919. This was the house journal of the Mouchel organisation.

26. The articles describing these were marked in Williams' own copies of this journal. Of particular interest appear to be those of Washington University, which was among a number that based their seating on earth construction (see 'Build large earth-fill

27. stadium by sheerboard method', *Engineering News Record*, 86, 1921, pp 326–29) and Kansas which adopted an arched arrangement for the seating supports rather like the Coliseum (see Williams C.C., 'Design features of the University of Kansas Stadium', *Engineering News Record*, 89, 1922, pp 790–93.). In the event, Wembley was to use both of these devices.

27. Figure from *Concrete and Constructional Engineering*.

28. 'The British Empire Exhibition.' *Concrete and Constructional Engineering*, 17, 1922, pp 699–703.

29. Williams O. 'The construction of the Empire Exhibition.' *Concrete and Constructional Engineering*, 19, 1924, pp 421–32. See also Blomfield R.T. (1932). *Memoirs of an Architect*. London, p188. Blomfield wrote of this scheme: *'I have profound admiration for engineers when they are dealing with steel and reinforced concrete but very little when they are dealing with bricks and mortar'*.

30. *Concrete and Constructional Engineering*, 8, 1924, p 421.

31. In Williams' own 1921 and 1922 copies of this journal all the articles on stadium design have been marked by him.

32. 'Build large earth-fill stadium by sheerboard method.' *Engineering News Record*, 86, 1921, pp 326–29.

33. 'Ohio Stadium, a double deck steel and concrete horseshoe.' *Engineering News Record*, 89, 1922, pp 640–44.

34. Barnes H. 'The British Empire Exhibition, Wembley.' *Architectural Review*, 55, 1924, pp 204–17.

35. Weaver L. 'Exhibitors' architecture.' *Architectural Review*, 55, 1924, pp 222.

36. Harvey W. 'The British Empire Exhibition and its concrete buildings.' *Concrete and Constructional Engineering*, 19, 1924, p 414.

37. Faber O. 'The concrete buildings.' *Architectural Review*, 55, 1924, pp 218-21.

38. Pite B. 'The architecture of concrete.' *Journal of the Royal Institute of British Architects*, 32, 1925, pp 329–36. Discussion pp 336-40.

39. Goodhart-Rendel H. S. (1953). *English architecture since the Regency*. Constable, London, p 259. The chapter from which this comes, 'The preferment of engineering', was originally delivered as a lecture in 1935. This was not the only aspect of concrete that he disapproved of because he (p 265) includes a reasoned argument against the use of set-back columns that had become popular as a means of obtaining horizontal emphasis in elevations.

40. *Architects Journal*, 64, 24 Nov,. 1926, pp 621–22

41. Bennet T. P. and Yerbury F. R. (1927). *Architectural Design in Concrete*. Benn, London. The photographs that formed the bulk of this book were captioned in both in English and German, perhaps suggesting that greater interest in the subject was expected from a German readership.

42. Ayrton, M. 'A note on concrete buildings.' *Journal of the Royal Institute of British Architects*, 31, 1924, pp 298–302.

43. Ayrton, M. 'A note on concrete buildings.' *Journal of the Royal Institute of British Architects*, 31, 1924, footnote. 2.

44. *Concrete and Constructional Engineering*, 18, 1924, p 421. Abstract of Williams' paper to the Institution of Municipal and County Engineers.

45. *British Engineers Export Journal*, July 1924.

46. Goodhart-Rendel H. S. (1953), *English architecture since the Regency*. Constable, London, p 259.

47. Discussion of Beresford Pite's lecture to the RIBA. *Journal of the Royal Institute of British Architects*, 32, 1925, p 338.

48. Ayrton M. 'A note on concrete buildings.' *Journal of the Royal Institute of British Architects*, 31, 1924, p 298.

49. Ayrton M. 'A note on concrete buildings.' *Journal of the Royal Institute of British Architects*, 31, 1924, p 298.

50. Williams' Papers, Birmingham. Unpublished manuscript, 'The economic proportioning of reinforced concrete construction'.

Chapter Three

51. Ayrton M. 'Modern Bridges.' *Journal of the Royal Institute of British Architects*, 38, 1931, p 487.

Endnotes

52. Barman C. 'A model partnership.' *Architects Journal*, 64, 1926, p 646.

53. See Williams O. 'The effective and the efficient.' *The Studio*, 101, 1931, pp 79–83.

54. See 'The Great North Road over the Grampians.' *Minutes of Proceedings of the Institution of Civil Engineers*, 232, 1930–31, pp 113–130.

55. Pite, B. 'The aesthetics of concrete.' *Journal of the Royal Institute of British Architects*, 32, 1925, pp 329–40.

56. Pite, B. 'The aesthetics of concrete.' *Journal of the Royal Institute of British Architects*, 32, 1925, pp 329–40.

57. Pite, B. 'The aesthetics of concrete.' *Journal of the Royal Institute of British Architects*, 32, 1925, pp 329–40.

58. 'Recent bridges by Sir Owen Williams.' *Concrete and Constructional Engineering*, 24, 1929, pp 281–88.

59. *Engineering News Record*, 91, 1923, pp 586–590. The article describing this bridge (and marked in Williams' bound copy of the journal) is also notable for another reason. The structure was a railway bridge of several large spans. Not only were the arches designed to have no tensile stresses but it was also an early practical trial of Duff Abrams' methods of proportioning concrete mixes.

61. At that time the costs of road transport would have been a more significant factor than they are today. The price of manufactured building products was commonly adjusted for delivery costs.

62. 'Recent bridges by Sir Owen Williams.' *Concrete and Constructional Engineering*, 24, 1929, pp 281–88.

63. Wms. Drgs.261/1-12; 328/1-4; *Concrete and Constructional Engineering*, 24, 1929, pp 285–87.

64. Wms. Drgs.261/1-12; 328/1-4; *Concrete and Constructional Engineering*, 24, 1929, pp 285–87.

65. 'Sir Owen Williams' New Bridge at Montrose.' *Architect and Building News*, 125, 1931, p 305.

66. *Engineering News Record*, 107, 1931, pp 300, 540, 784; 108, 1932, pp 143 and 375. One of the American engineers called it a 'freak cantilever'.

67. Williams is here responding to Professor Beresford Pite's lecture.

68. 'Beam and slab concrete highway bridges to carry Ministry of Transport loadings.' *Journal of the Institute of Municipal and County Engineers*, 52, 1926, pp 985–91.

69. Williams O. 'The philosophy of masonry arches.' *Selected Engineering Papers*, 56 (1927), The Institution of Civil Engineers, London.

70. 'Towards simplicity.' *Concrete and Constructional Engineering*, 21, 1926, p 19.

71. *Daily Express*, 12 October 1932.

72. *Evening Standard*, 12 October 1932.

73. *Concrete and Constructional Engineering*, 27, 1932, pp 281–83.

74. *Concrete and Constructional Engineering*, 27, 1932, pp 281–83.

75. *Architects' Journal*, 81, 1935, p 900. Included a model of the proposed bridge.

76. Shand M. 'Concrete and Steel.' *Architectural Review*, 72, 1932, p 176.

Chapter Four

77. Mendelsohn E, *Structures and sketches translated from the German by H. G. Scheffauer* (1924), Benn, London.

78. Ayrton M. 'Surface, true and false.' *Architects Journal*, 64, 1926, pp 661–64.

79. *Architect and Building News*, 20, 1929, pp 346–47.

80. *Architects Journal*, 82, 1931, pp 105–07; *Concrete and Constructional Engineering*, 24, 1929, pp 459–66.

81. Brockman H. A. N. (1974). *The British Architect in Industry, 1841–1940*. London, pp 150–51.

82. Bennet T. P. and Yerbury F. R. (1927). *Architectural Design in Concrete*.

83. *Architect and Building News*, 117, 1927, p 444.

84. *Journal of the Royal Institute of British*

Architects, 38, 1931, p 493. A response to Ayrton's paper, 'Modern Bridges'.

85. In a debate at the Architecture Club he proposed the motion 'That there is no fundamental difference between Architecture and Engineering'. This was opposed by Goodhart-Rendel and Howard Robertson. However, Williams seems to have set up a semantic hedge that largely frustrated sensible debate. *The Architect and Building News*, 125, 1931, p 200.

86. Richards J. M. (1980). *Memoirs of an Unjust Fella*. Weidenfeld and Nicholson, London, p 50.

87. Bromley Watkins Crawley and Bros.

88. *The Observer*, 23 Feb. 1930.

89. Lutyens had been used as a consultant for Grosvenor House, an adjoining buildng on Park Lane. He was largely responsible for the façades with the planning undertaken by Ewimperis, Simpson and Gutheris. See *Architect and Building News*, 119, 1928, pp 887–89.

90. McAlpines were also contractors for the Duntocher, Findhorn, Spey and Montrose bridges.

91. *The Yorkshire Post*, 28 Nov. 1929.

92. *The Daily Telegraph*, 20 Dec. 1929.

93. See 'Northcliffe House, London.' *The Builder*, 135, 1929, p 309; 'Northcliffe House, Bristol.' *The Builder*, 136, 1929, p 687; *Architect and Building News*, 11 Oct. 1929, plate 35; *The Builder*, 137, 1930, p 994; *Architect and Building News*, 149, 1936, pp 71–74.

94. Published in Farey C.A. and Trystan Edwards A. (1931). *Architectural Drawing, Perspective and Rendering*, London, and reproduced in a review of this book by *Architectural Review*, 69, 1931, p 211.

95. Based on anecdotal evidence from an interview with R. E. Foot, partner of Sir Owen Williams and Partners.

Chapter Five

96. *The Daily Telegraph*, 27 Nov. 1929.

97. Towndrow F. 'The engineer in Park Lane, Beauty and efficiency.' *The Observer*, 1 Dec. 1929.

98. Given Towndrow's favourable comments on Williams' appointment, it is surprising that he made no mention of him or his work in his book *Architecture in the Balance*, Chatto and Windus, London, 1933.

99. The story of J. M. Richards transfer to Williams' office is told by him in 'A brace of original hotels', *The Listener*, 11 Sept. 1969, p 34 and also in *Memoirs of an Unjust Fella* (1980), p 49.

100. 'Concrete in Park Lane.' *The Observer*, 23 Feb. 1930.

101. Towndrow F. 'The engineer in Park Lane, Beauty and efficiency.' *The Observer*, 1 Dec. 1929.

102. *Architects Journal*, 71, 1930, p 3.

103. Anecdote by O.T. Williams when interviewed by David Cottam in Febuary 1982.

104. 'New sensation over giant concrete hotel — Sir Owen Williams Resigns.' *Daily Chronicle*, 8 March 1930.

105. The story of the taxi is told by Richards in *Memoirs of an Unjust Fella* (1980), p 51.

106. Dorchester file photographs — Williams' Papers, Birmingham.

107. 'The Dorchester Hotel.' *Concrete and Constructional Engineering*, 26, 1931, 289–303; *Architect and Building News*, 125, 1931, pp 105–124; *Architects Journal*, 73, 1931, pp 577–82. See also Curtis Green's drawings (Green Lloyd and Adams Office, London).

108. *Architect and Building News*, 125, 1931, pp 105-24.

109. Robertson H. (1932). *Modern Architectural Design*. Architectural Press, London, pp 14–15. He uses the Dorchester to illustrate how framed structures were liberating the plan, wrongly crediting it to Curtis Green.

110. Leathart J.(1940). *Style in Architecture*. Thomas Nelson, London, pp 150–51.

111. *Architect and Building News*, 125, 1931, pp 105–24.

112. *The Listener*, 11 Sept. 1969, p 34.

113. Chermayeff S. 'The new building for the Daily Express.' *Architectural Review*, 72, 1932, pp 3–12.

Endnotes

114. 'The Daily Express Building, Fleet Street.' *Architect and Building News*, 131, 1932, 11–17.

115. *The Builder*, 15 July 1932, p 88.

116. Williams' accounts correspondence file A–D.

117. *Glass*, 9, 1932, p 353. They also began to display their products at the Building Exhibition in that year.

118. Reilly C. 'The year's work.' *Architects Journal*, 77, 1933, p 48.

119. Goodhart-Rendel H. S. (1953). *English Architecture Since the Regency*, Constable, London, pp 258–59. Originally a lecture given in 1935, 'The preferment of engineering'.

120. Chermayeff S. 'Daily Express.' *Architectural Review*, 72, 1932, pp 3–12.

121. *Concrete and Constructional Engineering*, 26, 1931, pp 289–303.

121a. Remarks made during a lecture to the Architectural Association on the design of the Empire Pool. *The Builder*, 28 May 1935.

122. *The Times*, 27 March 1956. In reply to a statement in the paper of 21 March 1956, which referred to Curtis Green and Partners as the architects of the Dorchester Hotel.

Chapter Six

123. *Architect and Building News*, 130, 1932, p 56.

124. See, for example, Richards J. M. (1940, 1962 2nd edn). *An Introduction to Modern Architecture*. Penguin, Harmondsworth, p 85. Banham R. (1962). *Guide to Modern Architecture*, Architectural Press, London, pp 169 and 170.

125. Williams E. O. 'Factories — a few observations thereon made by Sir Owen Williams at a discussion of the Art Workers Guild, 21 October, 1927.' *Journal of the Royal Institute of British Architects*, 35, 1927, pp 54–55. The lecture was originally to have been given by Thomas Wallis. Williams replaced him due to illness.

126. Walker E. (1977). *100 Years Shopping at Boots*. Boots, Nottingham.

127. Kahn M. (1917). *The Design and Construction of Industrial Buildings*. Technical Journals, London.

128. *Architects Journal*, 64, 1926, p 715.

129. 'Special features of a concrete building.' *Engineering News Record*, 71, 1914, p 602.

130. Boots have been unable to floor these atria to overcome this problem because of the building's listed status.

131. Its introduction to Britain has been described in more detail by David Yeomans (1997) in *Construction since 1900: Materials*. Batsford, London, pp 121–24.

132. These treated the flat slab as a series of cantilevers for the purposes of calculation and it could not be fitted into such a model.

133. *Concrete and Constructional Engineering*, 21, 1926, p 28; and, 22, 1927, pp 112–22.

134. *Concrete and Constructional Engineering*, 21, 1926, p 28; and, 22, 1927, p 727.

135. *Architectural Review*, Sept. 1933, pp 91–93.

136. Korn A. (1929). *Glas im Bau und als Gebracuchgegenstand*. Pollak, Berlin.

137. Crabb G. 'Life in Sir Owen Williams' office, 1927–38', in Cottam D. (1986). *Sir Owen Williams, 1890–1969*. Architectural Association, London, p 147.

138. In 'The year's work', *Architects Journal*, 77, 1933, p 48. Reilly had not actually visited the building, his report making it clear that he was working from photographs of it.

139. Brockman H. A. N. (1974). *The British Architect in Industry, 1841–1940*. Allen and Unwin, London, pp 150–51.

140. *Building*, 7, 1932, p 392.

141. 'New Piccadilly Circus Garage.' *Architect and Building News*, 121, 1929, pp 475–77. A number of garage structures were reported in that year.

142. Background information can be found in Pearse I .H. and Crocker L. H. (1943). *The Peckham Experiment – A Study in the Living Structure of Society*. Allen and Unwin, London.

143. Pearse I. H. and Williamson G. S. (1930). *The Case for Action*. Scottish Academic Press, London.

144. Exhibited at the Royal Academy in 1931. Events suggest that the principal

purpose of this was as an advertisement to attract funds.

145. Richardson J. M. (1980). *Memoirs of an Unjust Fella.* Weidenfeld and Nicolson, London, pp 88–90.

146. The RIBA library, GreH/3/1/1-53.

147. The RIBA library, GreH/3/1/1-53. It is apparent from the archives that a number of architects did not share Goodhart-Rendel's scruples and did indeed submit prices.

147a. Letter to the RIBA. The RIBA library, GreH/3/1/1-53.

148. Richards J. M. 'The Pioneer Health centre' and 'The Idea Behind the Idea'. *Architectural Review,* 77, 1935, pp 203–16.

149. In his *Architectural Review* article Richards went to some length to show that the building, while appearing symmetrical, was in fact asymmetrical. However, there is little doubt that the building is based upon a symmetrical axial arrangement. For example, at the second floor level Williams included two identical spaces on the southeast and northwest façades labelled 'study and recreation.' This appears to have been to preserve the symmetry, for it is unlikely that two identical spaces of this size were needed.

150. Perspective from Williams' Papers, Birmingham.

151. Pearse I.H. and Crocker L. H. (1943). *The Peckham Experiment — A Study in the Living Structure of Society.* Allen and Unwin, London, p 68.

Chapter Seven

152. *Architect and Building News,* 3 Nov. 1933, p 125.

153. Perlmutter R. and Mark T. 'Engineer's Aesthetic vs architecture, the design and performance of the Empire Pool at Wembley.' *Journal of the Society of Architectural Historians,* 31, 1972, pp 56–60. Their suggestion that the span of the building was designed to be a complete number of units, 86, seems less likely, especially as the gridlines drawn on the plan do not correspond with the long walls. See also the report on Williams' lecture to the Architectural Association, *The Builder,* 31 May 1935, p 1026.

154. The main building was accompanied by a small solvent recovery building but this need not be discussed.

155. *Architect and Building News,* 26 Feb. 1937, p 265.

156. *Architectural Association Papers,* Files 12. This was a review of the Architectural Association exhibition on the work of Owen Williams in 1986.

Chapter Eight

157. Sir Allan Harris, review of the Owen Williams exhibition at the Architectural Association. *Architectural Association Papers,* Files12.

158. Reilly C. 'The year's work.' *Architects Journal,* 85, 1937, p 100. The same journal had described the building previously with a number of photos of the inside, see *Architects Journal,* 84, 1936, pp 234–37.

159. *Architect and Building News,* 147, 1936, pp 161–64.

160. *Architect and Building News,* 150, 1937, pp 22–24.

161. Faber O. 'The engineer as designer.' *The Structural Engineer,* 12, 1934, p 73.

Chapter Nine

162. Williams O. 'The Portent of Concrete.' *Concrete and Constructional Engineering,* 27, 1932, pp 42–43. Also, *Architects Journal,* 74, 1931, p 829.

163. Faber O. 'The Portent of Concrete — an answer to Sir Owen Williams.' *Architects Journal,* 76, 1932, pp 121–22.

164. Faber O. 'The Portent of Concrete — an answer to Sir Owen Williams.' *Architects Journal,* 76, 1932, pp 121–22.

165. *Architectural Review,* 72, Nov. 1932, p 162.

166. Trystan Edwards A. 'The structural engineer as artist.' *The Structural Engineer,* 4, 1926, p 25 et seq.

167. *Evening News,* 15 Feb. 1930.

168. *The Daily Telegraph,* 7 March 1930.

169. Williams L. 'Building with glass.' *Architectural Review,* 73, 1933, pp 92–93.

170. *Architectural Review,* 72, Nov. 1932.

171. *Architects Journal,* 81, 1935, pp 561–62.

Selected list of works

1921-24
British Empire Exhibition for British Empire Exhibition incorporated
Architects, Simpson and Ayrton

1924-25
Lea Valley Viaduct & bridge for Ministry of Transport
Consulting architect, Maxwell Ayrton

1924-25
Park des Attractions, Paris

1924-26
Findhorn Bridge for Ministry of Transport
Consulting architect, Maxwell Ayrton

1924-27
Bridge, Shepherd Leys Wood, Kent, for Ministry of Transport.

1924-29
Bournemouth Pavilion, Hampshire, for Bournemouth Pavilion Committee
Architects, Horne & Knight

1925-26
Spey Bridge for Ministry of Transport
Consulting architect, Maxwell Ayrton

1925-26
Crubenmore & Loch Alvie Bridges for Ministry of Transport
Consulting architect, Maxwell Ayrton

1925-26
Duntocher Bridge, Glasgow, for Glasgow Corporation
Consulting architect, Maxwell Ayrton

1925-26
Belfast Water Tower, for Sir Robert McAlpine & Sons

1925-28
Wansford Bridge for Ministry of Transport
Consulting architect, Maxwell Ayrton

1926-28
Dalnamein Bridge for Ministry of Transport
Consulting architect, Maxwell Ayrton

1926-28
Carr Bridge for Ministry of Transport (demolished)
Consulting architect, Maxwell Ayrton

1926-28
Lochy Bridge for Inverness County Council and Ministry of Transport
Consulting architect, Maxwell Ayrton

1927-28
Brora Bridge for Sutherland County Council
Consulting architect, Maxwell Ayrton

1927-30
Montrose Bridge for the Royal Borough of Montrose
Consulting architect, Maxwell Ayrton

1928
Exhibition Hall, Salford for Laycock & Bird.
Architects, Robert Atkinson

1928-29
Pont-Rhyd-Owen Bridge for Carmarthen County Council

1928-30
Wadham Rd. Viaduct for Ministry of Transport
Consulting architect, Maxwell Ayrton

1928-30
Harnham Bridge, Salisbury, for Wiltshire County Council

1928-30
Steyning Farm, Sussex for the Hon Arthur Howard
Architect, Maxwell Ayrton

1928-30
Pilkington's Warehouse, London, for Mssrs Pilkington Bros (demolished)
Architect, Maxwell Ayrton

1929-30
Dorchester Hotel. Completed by Curtis Green with Considère and Partners

1929-31
Llechryd Bridge, (unexecuted) for the Counties of Cardigan and Pembroke

1929-31
Wakefield Bridge for the City of Wakefield

1929-31
Daily Express Building, Fleet Street for the Daily Express Building Company
Architects, Ellis & Clarke

1930-32
Boots Wets Building, Beeston, Nottingham for Boots Pure Drugs Company

1931-33
Sainsbury's factory and warehouse, Blackfriars, London for James Sainsbury

1932
Waterloo Bridge

1932-33
Tunnel Cement Laboratory, Thurrock Essex for Tunnel Portland Cement Company

1932-34
Cumberland Garage and car park, Marble Arch London for J Lyons & Co

1933-34
Empire Pool, Wembley, for Wembley Stadium Ltd
1933-35
Pioneer Health Centre, Peckham, for Pioneer Health Centre Ltd. (subsequently altered by the LCC)
1933-36
Residential flats, Stanmore for Owen Williams (Basic Buildings Ltd.)
1935-37
Provincial Newspaper Offices, London for Provincial Newspapers Ltd
1935-38
Boots factory extensions for Boots Pure Drugs Company
1935-38
Odham's printing works, Watford for Odhams Press Ltd
1935-39
Daily Express, Manchester for Beaverbrook Associated Newspapers Ltd
1936-37
Lilley & Skinner office and warehouse extension for Lilley & Skinner Ltd
1936-38
Dollis Hill Synagogue, Cricklewood, London, for United Synagogue
1936-39
Scottish Daily Express Building, Glasgow, for Beaverbrook Newspapers Ltd
1937-38
Removable restaurant, Wembley Stadium for Wembley Stadium Ltd

1937-39
Tunnel Cement Laboratory and Offices, Pitstone, Bucks. For Tunnel Portland Cement Company
1937-39
Hunt Partners extension, Clapton, London for Hunt Partners
1938-39
Daily News Garage, Southwark for Daily News Ltd
1939-41
Vickers Armstrong Aircraft Factory, Blackpool - Completed by Oscar Faber and Partners
1945-67
Newport By-pass for the Ministry of Transport Consulting architects, Sir Percy Thomas & Son
1950-55
BOAC Maintenance Headquarters, Heathrow for the Ministry of Civil Aviation
1951-67
M1 Motorway - carried out in two phases for the Ministry of Transport
1953-66
Port Talbot Bypass, West Glamorgan for the Ministry of Transport
1954-56
BOAC Wing hangars, Heathrow for the Ministry of Civil Aviation
1955-61
Daily Mirror Building, Holborn for Daily Mirror Newspapers Ltd
Associated architects, Anderson, Forster & Wilcox
1959-60
Daily Express, Manchester – extension for Beaverbrook Newspapers Ltd

Bibliography

Works by Owen Williams

1919 - 'The economic size of concrete ships', *Engineering*, 107, 195-97.
1924 - 'The construction of the Empire Exhibition', *Concrete and Constructional Engineering*, 19, 421-32.
1924 - 'Concrete as a partnership of engineering and Architecture', *British Engineers Export Journal*, July.
1924 - 'Brick for Houses', *Journal of the RIBA* 6 Dec.
1925 - 'The relative economy of different classes of concrete for reinforced concrete work', *Engineering*, 27 November.
1926 - 'Towards simplicity', *Concrete and Constructional Engineering*, 21, 19.
1926 - 'Beam and slab concrete highway bridges to carry Ministry of Transport loadings', *Journal of the Institute of Municipal and County Engineers*, 52, 985-91.
1927 - 'The Philosophy of Masonry Arches', *Selected Engineering Papers*, no. 56, The Institution of Civil Engineers, London.
1927 - 'Factories', *Journal of the RIBA* 35, 54-55.
1928 - 'Bridges', *Journal of the RIBA*, 35, 296-304.
1930 - 'The law of least action as a basis of engineering and architecture', *Aberdeen Press* 12 February.
1930 - 'Rapid-hardening Portland cement' *Concrete and Constructional Engineering*, 25, 80-1.
1931 - 'The effective and the efficient', *The Studio*, 101, 79-83.
1932, 'The portent of concrete', *Concrete and Constructional Engineering*, 27, 42-43. Also *Architects Journal*, 74.
1932 - 'A concrete thought', *Architectural Review*, 72, 162.
1933 - 'Modern factory design in England', *Engineering News Record*, 111, 675-76.
1952 - 'Architecture - Trade Profession or calling', *Architectural Association Journal*, December, 670-72.
1956 - 'The motorway and its environment', *Architects Journal*, 123, 98-105.

Works by others

Ayrton, M. (1924), 'A note on concrete buildings', *Journal of the Royal Institute of British Architects*, 31, 298-302.
Ayrton, M. (1926), 'Surface, true and false', *Architects Journal*, 64, 661-64.
Ayrton, M. (1931), 'Modern Bridges', *Journal of the Royal Institute of British Architects*, 38, 479-95.
Banham, R (1962), *Guide to Modern Architecture*, Architectural Press London.
Barman, Christian (1926), 'A model partnership', *Architects Journal*, 64, 646-51.
Barnes, H. (1924), 'The British Empire Exhibition, Wembley', *Architectural Review*, 55, 207-17.
Bennet, T. P. and F. R. Yerbury (1927), *Architectural Design in Concrete*, London, Ernest Benn.
Black, William (1914), 'The architectural treatment of reinforced concrete', *Kahncrete Engineering*, 1, 33.
Blomfield, A. (1871), 'The use of Portland cement concrete as a building material' *RIBA Sessional Papers*.
Blomfield R. T. (1932), *Memoirs of an Architect*, McMillan, London.
Brockman, H. A. N. (1974), *The British Architect in Industry, 1841-1940*, Allen and Unwin: London.
Cottam, David (1986), *Sir Owen Williams, 1890-1969*, Architectural Association: London.
Chermayeff, Serge (1932), 'The new building for the Daily Express', *Architectural Review*, 72, 3-12.
Collins, Peter (1959), *Concrete, The Vision of a New Architecture: A Study of Auguste Perret and his Predecessors*, Faber & Faber, London.
Edwards, A. Trystan (1926), 'The structural engineer as artist', *The Structural Engineer*, 4, 60-64, 84-87, 126-30, 155-58, 191-95, 221-24, 251-54, 283-86, 341-44, 375-79.
Faber, Oscar (1924), 'The concrete buildings', *Architectural Review*, 55, 218-21.
Faber, Oscar (1932), 'The portent of concrete – an answer to Sir Owen Williams', *Architects Journal*, 76, 121-2.
Faber, Oscar (1934), 'The engineer as designer', *The Structural Engineer*, 12, 73.
Farey, C. A. and A. Trystan Edwards (1931), *Architectural Drawing, Perspective and Rendering*, Batsford, London.

Goodhart-Rendel, H. S. (1953), *English Architecture since the Regency*, Constable, London.

Harvey, William (1924), 'The British Empire Exhibition and its concrete buildings', *Concrete and Constructional Engineering*, 19, 410-19.

Kahn, Moritz (1917), *The Design and Construction of Industrial Buildings*, London, Technical Journals.

Korn, A. (1929), *Glas im Bau und als Gebracuchgegenstand*, Ernst Pollak, Berlin.

Lethart, Julian (1940), *Style in Architecture*, Thomas Nelson, London.

Luckhurst, K. W. (1951), *The Story of Exhibitions*, Studio Publications, London.

Marsh, C. F. and W. Dunn (1906), *Reinforced Concrete*, Constable, London, 3rd ed.

Mendelsohn, Eric (1923), *Structures and Sketches translated from the German by H. G. Scheffauer*, Benn, London.

Pearse, I. H. and L. H. Crocker (1943), *The Peckham Experiment – A Study in the Living Structure of Society*, Allen and Unwin, London.

Perlmutter, Roy and Robert Mark (1972), 'Engineer's Aesthetic vs architecture, the design and performance of the Empire Pool at Wembley', *Journal of the Society of Architectural Historians*, 31, 56- 60.

Pite, Beresford (1925), 'The architecture of concrete'. *Journal of the Royal Institute of British Architects*, 32, 329-40.

Reilly, Charles (1933), 'Built now: The year's buildings', *Architects Journal*, 77, 36-47.

Reilly Charles (1937), 'The year's work at home', *Architects Journal*, 85. 91-102.

Richards J. M. (1935), 'The Pioneer Health centre' and 'The Idea Behind the Idea', *Architectural Review*, 77, 203-16.

Richards, J. M. (1940), *An Introduction to Modern Architecture*, Penguin, Harmondsworth.

Richards, J. M. (1969), 'A brace of original hotels', *The Listener*, 11 Sept, 34.

Richards, J. M. (1980), *Memoirs of an Unjust Fella*, Weidenfeld and Nicholson, London.

Robertson, H. (1932), *Modern Architectural Design*, Architectural Press, London.

Shand, P. Morton (1932), 'Concrete and Steel', *Architectural Review*, 72, 169-79.

Towndrow, F (1929), 'The Engineer in Park Lane, Beauty and Efficiency', *The Observer*, 1 Dec.

Towndrow, F. (1933), *Architecture in the Balance*, Chatto and Windus, London.

Walker, E (1977), *100 Years of Shopping at Boots*, Boots Co. Ltd., Nottingham.

Weaver, Lawrence (1924), 'Exhibitors' architecture', *Architectural Review*, 55, 222-29.

White, R. B. (1965), *Prefabrication: A history of its development in Great Britain*, HMSO, London.

Williams, Llewellyn (1933), 'Building with glass', *Architectural Review*, 73, 92-93.

Yeomans, D. (1997), *Construction since 1900: Materials*, Batsford, London.

Yeomans, D. (2000), 'Collaborating with consulting engineers', in Louise Campbell, ed. *Twentieth-Century Architecture and its Histories*, Society of Architectural Historians of Great Britain, London, 125-51.

Index

A9 bridges 36–8, 39
aesthetic theory 34–51
aircraft hangars 117–23
American Shredded Wheat Company 88
arch bridges 38-40, 42-3, 46-8, 51
Architect and Building News
    Boots factory 92
    Daily Express building 68, 70
    Dorchester Hotel 66
    double cantilever bridges 44-5
    Empire Pool 108, 144
    flat-slab construction 56, 57, 88, 95
    Lilley and Skinner 129
    minor buildings 137
    Odhams printing works 112-13
    Provincial Newspaper offices 133
Architects Journal 28, 54, 64, 86, 131
Architectural Association 109
Architectural Review
    Barnes, Harry 26
    Chermayeff, Serge 78, 143
    Daily Express building 68, 74
    Pioneer Health Centre 144
    Williams, Llewellyn 142
Art Workers' Guild 82
atrium spaces 85, 86, 91, 92

Barnes, Harry 26
Beaux-Arts strategy 20
Bedford College (London University) 55
Bennet, T. P. 12, 28, 56
Berliner Tageblatt 54
Bernard, Oliver 62, 63, 64

Birmabright 75
Blomfield, Arthur 14, 21
BOAC *see* The British Overseas Airways Corporation
Boots Company 57, 58, 88, 89, 92
    'Drys' building 105, 114-17, 145, 146
    fire station 132-3
    'Wets' building 82-7, 90, 114, 115, 137, 140, 146
bridges 34-51, 54, 56, 70, 104
British Empire Exhibition 16, 17, 20-31, 34, 140
British Overseas Airways Corporation 117-23
British Vitrolite 73, 74, 75, 76
Broadhead, F. A. 88
The Builder 21, 72
Building 92, 144

cantilevers
    Boots factory 89
    Daily Express building 72
    double cantilever bridges 44-6
    long-span buildings 116-23
    Pite, Prof Beresford 28
    Sainsbury factory 94
Centenary Hall (Breslau) 28
Chermayeff, Serge 68, 71, 78, 140, 143
Collins, Peter 21-2
Concrete 26
Concrete and Constructional Engineering 42, 43, 50, 88
Concrete Institute 12, 21
Concrete Utilities Bureau 26
Considère and Partners 65, 79
costs 141

Crabb, George 91
Cumberland garage and car park 94

Daily Express building 57, 58, 59, 62, 67-78
Daily Mirror building 135
Daily Telegraph 62
Dalnamein Bridge 43, 104
Dollis Hill synagogue 129-32, 145
Donaldson, J. G. S. 95
door structures 121-2
Dorchester Hotel 57, 58, 62-7, 79, 137, 143
double cantilever bridges 44-6
'Drys' building *see* Boots Company
Dunn, W. 12
Duntocher Bridge 43

'economy of means' principle 29-30
Edwards, Trystan 142
Ellis, Clough Williams 26
Empire Exhibition 16, 17, 20-31, 34, 140
Empire Pool building 108-12, 129, 144
Engineering News Record 14, 24, 26, 45

Faber, Oscar 13, 26, 137, 140, 141
fin structures 110-11
Findhorn Bridge (Tomatin) 36-8, 48
fire station (Boots) 132-3
flat-slab buildings 82-105, 115, 144, 145
Fleet Street Daily Express building 136
folded plate arrangements 129-32

General and Marine Concrete Construction (Williams System) 14

glazing	73-8, 90-2, 98-104, 112	Mark, T.	111	National Institute for Medical Research	
Goodhart-Rendel	28, 30, 77, 96, 97, 98	MARS *see* modern architectural research		(Hampstead)	55
Gramophone Company	58	mass concrete	42, 46-7	North Circular Road	34
Green, Curtis	64-8, 79, 143	Mendelsohn, Erich	54, 56		
		Metropolitan Tramway Company	12	The Observer	62, 63
H. O. Ellis and Clarke	58, 59	Middlesex County Council Building Regulations		Odhams printing works	112-14
hangars	17-23		109	Ohio Stadium	26
Harris, Allan	123	minor buildings	126-37	open spandrel arches	42-3
Harvey, William	26	Boots fire station	132-3		
hexagonal structures	118	Daily Mirror building	135	pavilions *see* British Empire Exhibition	
Horder, Morely	63, 64, 66	Dollis Hill synagogue	129-32	Pease, Innes H.	95
Hurst, B. L.	16	Lilley and Skinner	128-9	'pen' arrangement	118-19
		Pitstone laboratories	135	Perlmutter, R.	111
Indented Bar	14	Provincial Newspaper offices	133-4	Perret, Auguste	22, 86, 128
		Stanmore flats	127-8	Pilkington Glass	55-6, 77
Kahn, Moritz	14, 85	Thurrock laboratories	126-7	Pioneer Health Centre	95-103, 129, 144
Kahncrete Engineering	14, 21	*Modern Architectural Design*	67	Pite, Prof Beresford	21, 22, 28, 39-40, 46
Kavanagh, C. J.	49-50	modern architectural research (MARS)	142	Pitstone laboratories	135
		Modern Movement		'The Portent of Concrete' paper	141
Lang, Fritz	82, 86	Daily Express building	71	Provincial Newspaper building	133-4, 137
LCC *see* London County Council		discussion	137, 140, 142, 143, 144		
Le Corbusier	142	Dorchester Hotel	62	Reilly, Charles	77, 92
Lea Valley Viaduct	34-6, 48	flat-slab buildings	82	reinforcement	36, 45, 69-71, 112, 144, 145
Lilley and Skinner	128-9, 136	Pioneer Health Centre	97	Rendel *see* Goodhart-Rendel	
Lochy Bridge	44, 45, 48	Montrose Bridge	44, 47	RIBA *see* Royal Institute of British Architects	
London County Council (LCC) regulations	13,	Morandi, Ricardo	117	Richards, J. M.	62, 96, 97, 144
	74, 87, 141	Morrison, Herbert	49	Robert McAlpine and Sons	58
long-span buildings	108-23	Morton Shand, P.	50-1	Robertson, Howard	77
Lukenwalde hat factory	54	Murray, Keith	65	roof structures	91, 94, 113, 114, 121, 122
Lutyens, Sir Edwin	58, 143	'mushroom' construction	87-9, 104, 105, 144	Royal Institute of British Architects (RIBA)	
		Musman, E. B.	96, 99	Ayrton, M.	27, 31, 56
McAlpine, Sir Robert	58, 65	Myer-Kauffmann textile works	54	committee	12, 13
Manchester Daily Express building	104, 105, 136			Pioneer Health Centre	96, 97

159

Index

Pite, Prof Beresford	28
sessional paper	21
rusticated concrete	22-3
Sainsbury building (Blackfriars)	14, 93-4
Scotland	36-9, 40, 41, 42, 44
Scott, Sir Giles Gilbert	50
Scott Williamson, G.	95
Shepherd Leys Wood bridge	70
Shoe Lane *see* Daily Express building	
Simpson, Sir John	55
Small Arms Factory (Birmingham)	16
Spey bridge	38, 39, 41, 48
stair structures	86, 112
Stamford Stadium (Washington University)	26
Stanmore flats	127-8
steel v. concrete	141
Strathcona, Lord	20
structural efficiency	39-40
Style in Architecture	67
sun screens	135
Tanner, Sir William	13
Théâtre des Champs-Elysées	22
Thurrock laboratories	126-7
The Times	79
Towards a New Architecture	142
Towndrow, Frederic	62
The Trussed Concrete Steel Company of England	12, 13, 14, 16, 20
'unit principle'	85
'universal material'	142

Van Nelle factory (Rotterdam)	90, 92
Vitrolite	73, 74, 75, 76
Viyella factory (Nottingham)	88
Voysey, C. F. A.	22
Wallis Gilbert and Partners	55, 58
Wallis, Thomas	56, 58, 66, 74, 144
Wansford Bridge	40-2, 43
Waterloo Bridge	49-51, 54
Weaver, Lawrence	26
Wells Aviation Company	14
Wembley *see* British Empire Exhibition; Empire Pool Building	
'Wets' building *see* Boots Company	
Williams Concrete Structures Limited	16
Williams, Llewellyn	144, 145, 146
Williams System *see* General and Marine Concrete Construction	
wing hangars	122-3
Wrigley Company (Wembley)	55, 58
Yerbury, F. R.	12, 28, 56